Leckie
the education publisher for Scotland

Third Level
MATHS PROBLEM
SOLVING SKILLS
Benchmark edition

T0173586

Student Book

Keith Gordon,
Chris Pearce, Trevor Senior

© Leckie & Leckie Ltd 2020

001/29042020
ISBN 9780008406219
10 9 8 7 6 5 4 3 2 1

Published by
Leckie & Leckie
An imprint of HarperCollins Publishers
Westerhill Road, Bishopbriggs,
Glasgow
G64 2QT
T: 0844 576 8126 F: 0844 576 8131
leckiescotland@harpercollins.co.uk
www.leckiescotland.co.uk

Keith Gordon, Chris Pearce and Trevor Senior assert their moral rights to be identified as the authors of this work.

Project managed by Letitia Luff and
Project One Publishing Solutions Ltd
Edited and proofread by Joan Miller
Content design and typesetting by Linda Miles,
Lodestone Publishing Limited.
Layout by Planman Technologies India Pvt and Jouve
Illustrations by Jerry Fowler
Production by Therese Webb

Printed and bound by CPI Group (UK) Ltd, Croydon
CR0 4YY

A CIP Catalogue record for this book is available from the British Library.

The publishers wish to thank the following for permission to reproduce photographs:

Cover images – all © Thinkstock except for soccer ball © Comstock; gears, dice, protractor © Stockbyte

Pages 6, 8, 9, 11, 12, 14, 21, 24, 26, 27, 30, 32, 33, 37, 38, 41, 44, 47, 48, 52, 55, 57, 62, 66, 67, 69, 72, 76, 79, 82, 83, 85, 90, 92, 93, 94, 95, 96, 97, 100, 101, 102, 103, 104, 106, 107t, 107b, 108, 110, 112, 113, 114, 115, 116, 117 © istockphoto.com

Page 18 © Stockbyte
Page 36 © Thinkstock
Page 58 © photos.com
Map on page 20 © Collins Bartholomew Ltd

The publishers wish to thank the following for their permission to reproduce data:
page 110 www.direct.gov.uk; page 111 DEFRA www.defra.gov.uk; page 121 B&Q www.diy.com; page 125 www.scotrail.co.uk; page 126 Eurostar www.eurostar.com; pages 128–129 DVLA www.dvla.gov.uk; pages 131–133 Swiss Timing Ltd www.swiss-timing.com; pages 135–138 CIA www.cia.gov.

Every effort has been made to trace copyright holders and to obtain their permission for the use of copyright material. The authors and publishers will gladly receive any information allowing them to rectify any error or omission at the first opportunity.

Contents

Activities

Rich Tasks

Introduction

What this book is about

- Helping you understand how mathematics is used in society and how it has had an impact in the past and in the present.
- Showing you that mathematics can be used to help tackle problems you might come across in real-life, such as workplace problems, personal finance and activity planning.
- Helping you improve your confidence in mathematics, so you can work independently when you need to use mathematics in other aspects of your life outside the classroom.
- Encouraging you to apply the mathematical skills you already know in new situations.
- Developing your ability to interpret results and decide how successful you have been in your calculations.
- Giving you the chance to work with others and talk about what you are doing with mathematics.

What you will get out of it

- The chance to see how mathematics is embedded in all sorts of activities outside the classroom, and to use your mathematics skills in a range of real-life contexts.
- Opportunities to apply your mathematical skills to solve problems in creative and logical ways and to work with others in solving problems.
- An understanding of why mathematics is useful in life and work, and how you can use mathematics to help you with the important decisions needed for independent living.
- Greater confidence about your ability to use mathematics in real life.

Some things to think about

- People who can use mathematics confidently get the most out of life, learning and work. That could be you.
- Do not worry about making mistakes. Some of the tasks in this book have more than one answer, and there is almost always more than one method to get the answer. Sometimes we learn more from the mistakes we make than when we get everything right. The important thing is that you are able to think clearly about what you are doing and that you are able to explain how and why you are doing it.

- Mathematics is very useful. As you work through the problems in this book, think about similar problems in real life, such as working out the buses you might need to get to go to the cinema with friends, the games your team needs to win to become champions (or avoid relegation!) and how to work out how to fit your furniture into your bedroom.

Using this book

This book contains 39 activities, each with a number of tasks. The information you need to solve the problems is always provided within the activity and any associated data sheets but it's completely up you to work out how to use the information and what strategies to use to solve the problems.

The activities are graded in terms of difficulty. The first activities are 1-star problems and are an introduction to the development of problem-solving skills, using straightforward approaches in a familiar context. As you progress through the book, you will develop a greater range of problem-solving skills until you are able to work through the hardest 3-star problems. 3-star activities are multi-step problems which require a more creative approach.

Some activities have **extension tasks**. These are more challenging, and are a good way of discovering just how good your problem-solving skills are.

Data sheets

Some of the activities in this book use data sheets, which provide additional information needed for the problems. The data sheets are provided towards the end of the book on pages 118–138.

Answers

Answers for calculations are provided at the end of the book, on pages 139–144. Try not to look at the answers until you have completed each activity. Check your answers when you have completed an activity and if you have made a mistake, go through the problem again and work out where you made the mistake and what you need to do to avoid making the same mistake again.

Curriculum for Excellence

The Benchmarks for Numeracy and Mathematics published in 2017 highlights the importance of the following skills, the ability to:

- interpret questions
- select and communicate processes and solutions
- justify choice of strategy used
- link mathematical concepts
- use mathematical vocabulary and notation
- use mental agility
- reason algebraically
- determine the reasonableness of a solution.

Through these skills learners will be able to identify the strategies and processes required to solve the problems presented. We know through research that the best learning takes place when learners are actively solving problems and able to talk about their ideas. We also know that being able to solve problems in new and unfamiliar contexts is a strong indicator that learners have demonstrated mastery over that part of the curriculum.

Each activity in the book has identified the different experiences and outcomes from across Second, Third and Fourth Level. Also identified are links with other curriculum areas. This recognises that mathematics is an interconnected body of skills, knowledge and processes that are integral to many parts of the school curriculum, everyday life and industrial skills.

Endangered species

An endangered species is any species that is considered to be in danger of extinction. There are over 100 different animal species that are classed as endangered.

The table on the next page gives some data about a few endangered species. Use the information to answer the following questions, then move onto the tasks.

Warm-up questions

1   What is the **average normal lifespan** of a **whale**?

2   Do animals generally have **shorter or longer** lives in **captivity** than in the **wild**?

3   Approximately **how many lions** existed in Africa in **1950**?

4   Are **male chimpanzees** generally **heavier** or **lighter** than **female chimpanzees**?

5   You are told that a **grizzly bear** weighs **700 lbs**. Do you think it is **male or female**?

CfE Outcomes

I can display data in a clear way using a suitable scale, by choosing appropriately from an extended range of tables, charts, diagrams and graphs, making effective use of technology.
MTH 2-21a/MTH 3-21a

I can solve problems by carrying out calculations with a wide range of fractions, decimal fractions and percentages, using my answers to make comparisons and informed choices for real-life situations.
MNU 3-07a

Links with
Technology, Science, Social Studies

Animal	Population	Size	Lifespan
Bat	Some bat populations are counted in millions. Others are extremely low or declining.	Large bats can have a wingspan of 6 feet. Small bats can be less than an inch long.	Most bats live longer than most mammals of the same size. The longest known lifespan of a bat in the wild is about 30–40 years.
Grizzly bear	In 1950 there were about 50 000 grizzly bears in North America. Now there are about 1000 remaining, in five separate populations. In Alaska, there are over 30 000 grizzly bears.	**Height:** about 3–3$\frac{1}{2}$ feet at shoulders **Length:** 6–7 feet **Adult weight:** male 300–850 lbs; female 200–450 lbs	20–25 years
Whale	Varies with each species.	**Length:** varies, up to 110 feet **Weight:** varies, up to 150 tons	Whales normally live 20–40 years but they can live up to 80 years.
Lion	The lion population in Africa has reduced by half since the early 1950s. Today, fewer than 21 000 remain in all of Africa.	**Height:** males reach 4 feet, females are smaller **Length:** males reach 5-8 feet, females are smaller **Weight:** males reach 330–500 lbs, females weigh less	13 years, although they may live longer in captivity
Chimpanzee	An estimated 100 000 to 200 000 chimpanzees live in the wild.	**Height:** approximately 4 feet **Weight:** males 90–120 lbs, females 60–110 lbs	Chimpanzees rarely live past the age of 50 in the wild, but have been known to reach the age of 60 in captivity.

Task 1

Using the different **column headings**, sort the animals into any **order** that you think is appropriate. When you have **insufficient information**, you will need to make **decisions**. Consider whether there is a **link** between the **population** and **lifespan** of these animals.

Task 2

Prepare a **15-minute presentation** about **endangered species** for your mathematics class. Include **mathematical facts**. Use books or the internet to **research other endangered species** and then **present your findings**.

Task 3

You may need to use the **information** given in the **table** at the start of this activity to answer these questions.

1   What is the **maximum length** of a **whale**, in **metres**?

> 1 foot = 30 centimetres

2   A **lion** weighs **400 lbs**.

    How many **kilograms** does this lion weigh?

> 1 kg ≈ 2.2 lb

3   Work out the number of **grizzly bears** remaining altogether in the **five separate populations**, as a **percentage** of the **number in the US** in 1950.

4   **Bats** are the **slowest** mammals on Earth to reproduce. At **birth**, a **pup** weighs up to **25%** of its **mother's body weight**. If a **mother bat** weighs **8 grams**, how much could the **pup** weigh?

5   Before the current levels of whaling, the **humpback whale population** was **150 000**. The population has since **fallen** to **one-sixth** of this value. What is the **population now**?

6   The **DNA** of **chimpanzees** and **humans** are about **98.4% similar**. What percentage is not similar?

Football

The activity is based on the Scottish Football League.

Task 1

There are 10 teams in the Scottish First Division. In a season, every team plays every other team four times, twice at their own ground (**home**) and twice at the other team's ground (**away**). That means in each season, each team plays 36 matches.

If the team wins, they get **3 points**. If they draw, they get **1 point**. If they lose, they get **no points**. If two teams have the same number of points, then the team with the bigger **goal difference** will be placed higher.

> **Goal difference** is simply the number of goals scored **for** the team minus the number of goals scored **against** the team.

1   How many **matches** are played altogether in a **First Division season**?

2   A team has **scored 48 goals** and has **12 goals scored against it**. What is their **goal difference**?

3   A team has **scored 15 goals** and has a **goal difference of –8**. How many goals have been scored against the team?

4   Explain why, if the **goal differences for all 12 teams** are added up, the total will be **zero**.

5   If **after 10 matches** a team has **won 4 matches, drawn 3** and **lost 3**, how many **points** would they have?

6 What is the **maximum number of points** that a team could have **at the end of a season**?

7 In the 2010–11 season, Dunfermline Athletic only **lost 6 matches**.

　a What is the **minimum number** of points they could have had?

　b In fact, they had 70 points. Work out the **combinations of wins and draws** that would give this number of points over **36 matches**.

8 In the 2008–9 season, St Johnstone **won the league** with **65 points**. They **won 17 games**. How many matches did they draw and how many did they lose?

9 At the end of the 2010–11 season, the bottom club Stirling Albion had **20 points**. They lost **24** of their matches. What are the combinations of wins and draws which would give 20 points from the remaining matches?

10 This scattergraph shows the **points** and **goal differences** for all 10 teams in the First Division at the end of the 2009–10 season.

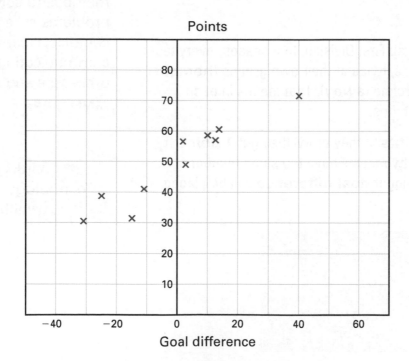

Read each of these statements and write down whether it is **true** or **false**.

　a The team with the most points had the greatest goal difference.

　b There must always be the same number of teams with a positive goal difference as there are with a negative goal difference.

　c The top team scored over 80 goals more than the bottom team.

　d If a team had a points total of 40, they would probably have had a goal difference of about 10.

Task 2

The Scottish FA Cup is a knock-out competition. It is has been played every year since 1873, except for during the two world wars. Entry is open to all 12 Premier League clubs, all 30 SFA League Clubs in Divisions 1, 2 and 3, and teams from three regional Junior leagues. The final is played at Hampden Park in Glasgow.

There are eight rounds in all. The first three rounds are qualifying rounds for teams from Leagues 1, 2 and 3, and from the Junior leagues. Premier League teams join at the fourth round stage. The final is the eighth round, so a Premier League team in the final plays in the fourth, fifth, sixth, seventh and eighth rounds. In the 2010-11 season, the first round was on 25 September 2010. The first round with Premier League teams was held on 8 January 2011 and the Final was held on 21 May between Celtic and Motherwell.

1 Two teams compete in the final, which is **round 8** of the competition. In **round 7**, (the semi-finals), there were **4 teams** competing. In **round 6**, (the quarter-finals), there were **8 teams** competing. How many teams competed in **round 4**?

2 In a Cup competition with **10 rounds**, how many teams would be in the **first round**?

3 For a Cup competition with *n* rounds, write down a rule for the number of teams that would be in the first round.

Task 3

The final score in a match is 3–2.

How many possible half-time scores are there?

Investigate this and come up with **a rule** for the number of possible half-time scores for any given final score.

1 Decide how to represent the problem, for example, in a diagram or a table, to make it easier to use mathematics to solve it.

2 Establish a pattern or relationship and then change the variables to see how this changes the results.

3 Test your generalisations and draw conclusions from the mathematical analysis.

4 Analyse the situation or problem and decide which is the most appropriate mathematical method that you would use to tackle it.

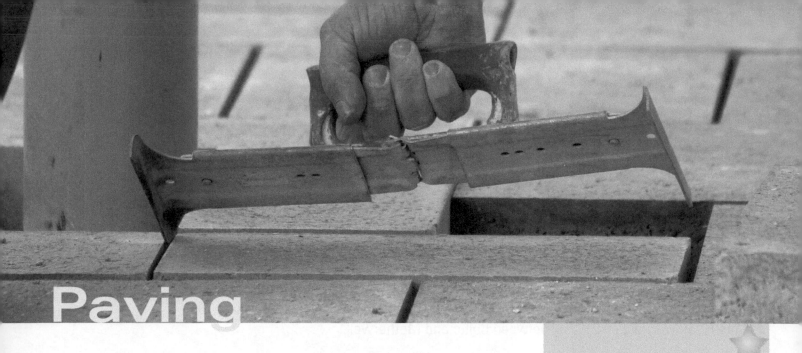

Paving

In this activity, suppose you are a builder. A customer has asked you to lay paving slabs to cover an area of her garden. You need to draw up the plans.

Garden: 60 metres x 10 metres
1 metre = 100 cm

60 m

10 m

Prices

- Square slabs cost £12 each or £99 for a pack of 10.

- Rectangular slabs cost £18 each or £120 for a pack of 10.

- Triangular slabs cost £6 each or £49 for a pack of 10.

20 cm
20 cm

30 cm
20 cm

20 cm

CfE Outcomes

I can manage money, compare costs from different retailers, and determine what I can afford to buy.
MNU 2-09a

I can solve practical problems by applying my knowledge of measure, choosing the appropriate units and degree of accuracy for the task and using a formula to calculate area or volume when required.
MNU 3-11a

Having investigated different routes to a solution, I can find the area of compound 2D shapes and the volume of compound 3D objects, applying my knowledge to solve practical problems.
MTH 3-11b

Warm-up questions

Use the data above to answer these questions.

1   How many centimetres are in a metre?

2   Four rectangular slabs are put in a line to make a path as shown.

How long is the path?

3   How much does it cost to buy 25 square slabs? Use the cheapest method.

4   How much does it cost to buy 25 square slabs and 31 rectangular slabs?
    Use the cheapest method.

5   What is the area of the garden to be paved, shown in the diagram? State the units in
    your answer.

6   How much more would it cost to buy 10 single triangular slabs than a pack of 10?

7   How many square slabs would fit in a square metre?

8   How many triangular slabs would cover the same area as 5 square slabs?

9   How many rectangular slabs could you buy for £100?

10 You have £540 to spend. Work out how many square, rectangular and triangular slabs
    you could buy if you buy the same number of each.

Task 1

Work out how many of the rectangular slabs would be needed to pave the whole garden shown
in the diagram.

Task 2

Now design a pattern of slabs to cover an area of the garden, using the square and
rectangular slabs.

Task 3

Work out the cost of your design for Task 2.

Task 4

Triangular slabs and square slabs can be fitted together as shown.
Design a pattern, using triangular and square slabs.
Use squared paper or graph paper.

Task 5 (extension)

Either design your own garden or design an L-shaped garden.
Use squared paper or graph paper.
Decide on the dimensions of your garden.
Remember to work out the cost of the slabs for your garden.
You may decide to add extra features to your design.

Money matters 1: Pay

Most people with a job get paid either a wage or a salary.

Most unskilled and semi-skilled workers, such as people who work in factories, stores or offices, are paid weekly by the hour and are known as wage-earners.

Most professionals, such as lawyers, doctors or teachers, are paid by the month and are known as salaried workers.

Wage-earners

There is no maximum for the amount a worker can earn an hour but there is a legally enforced national minimum wage.

There are three levels of minimum wage, and the rates from 1 October 2011 are:

- £6.08 per hour for workers aged 21 years and over
- a development rate of £4.98 per hour for workers aged 18–20 inclusive
- £3.68 per hour for all workers under the age of 18, who are no longer of compulsory school age.

Students at school who do a paper round, for example, can be paid below £3.68 per hour.

Task 1

Copy and complete the table for these workers.

NB: MW means they are paid minimum wage.

Name	Age	Hours	Wage per hour	Total weekly wages
Aftab	30	35	£7.20	252.00
Betty	25	40	MW	243.20
Colin	42	38	£13.50	513.00
Dierdre	17	30	MW	110.40
Eddy	28	35	£8.75	£306.25
Frank	21	31.4	MW	£243.20
Gus	15	8	2.75	£22.00
Hinna	25	37.5	£9.80	£367.50
Ian	18–20	36	MW	£179.28
Jemima	<18	20	MW	£73.60

Overtime

Most wage-earners have a **fixed number of hours** that they have to work each week, typically around 40 hours. If they are required to work beyond these hours then they will get paid **overtime**. The pay rate for overtime will be **more than** the normal rate. For example, overtime may be paid at **time-and-a-half**, which means that a worker normally earning £8 per hour would get £12 per hour for overtime paid at time-and-a-half.

Task 2

Copy and complete this table.

NB: MW means they are paid minimum wage.

Name	Age	Basic hours	Overtime hours	Basic wage per hour	Overtime rate	Total wages
Alf	32	35	10	£8.20	1.5 x	
Belinda	27	40	6	MW	1.25 x	
Chas	42	36	8	£11.50	1.5 x	
Dave	17	35	5	MW	2 x	
Edith	29	40		£8.75	1.4 x	£411.25
Francis	20	35		MW	1.5 x	£204.18
Gaynor	14	10	2		1.5 x	£41.60
Henry	26	40	6		1.25 x	£427.50
Iris		38	8	MW	1.5 x	£249.00
Jack		35	10	MW	1.25 x	£174.80

Salaried staff

Workers who are paid a **salary** get the **same pay each month**, even though some months are shorter than others. This is calculated by **dividing the annual salary by 12**.

Salaried workers do not usually work overtime, although they may have a **fixed number of hours**. If they work beyond the required number of hours they may be allowed a **day off to compensate** or **receive a bonus payment**.

Task 3

Copy and complete this table.

Name	Monthly salary	Annual salary
Pete	£2500	
Quinlan	£3350	
Rosie		£22 500
Sue		£74 520
Teresa	£4180	

Extension questions

1   Arnold works a basic **35-hour week** for a basic wage of **£9.50 per hour**. One week he works his basic hours plus **8 hours'** overtime at **time-and-a-half** and another **8 hours'** overtime at **double time**. What is Arnold's **total wage**?

2   Bettina, 20, who is on **minimum wage**, works a basic week of **40 hours** from Monday to Friday. She does an **extra hour each day** from **Monday to Friday** at **time-and-a-quarter** and then works **8 hours** on Sunday for **time-and-a-half**. What is Bettina's **total wage** for that week?

3   Colleen works a basic week of **38 hours**. She also works **6 hours'** overtime at **time-and-a-half** and **4 hours'** overtime at **double time**. Her **total wage** for that week is **£412.50**. What is Colleen's **hourly rate**?

4   Derek is paid **minimum wage** and works a basic week of **40 hours**. He also does **6 hours'** overtime at **time-and-a-quarter** and the **same number of hours'** overtime at **time-and-a-half**. His total wage is **£281.37**. How **old** is Derek?

5   Erica works **8 hours** each day from Monday to Friday. She is paid **minimum wage**. **Wednesday** last week was her **21st birthday**. What was her **total wage** for that week?

Scotland

In this activity, you will find out about more about Scotland. You will need to refer to the **Data sheet: Scotland** (pp 118–119).

Task 1

Use the information on the **Data sheet: Scotland** to answer these questions.

1   What **percentage** of **people living in Scotland** in **2001** were **born** outside Scotland?

2   Which **region** of Scotland saw the **greatest population growth** between **1861** and **1911**?

3   How many **kilometres** of **coastline** are there in Scotland?

4   What is the **height** of the **highest mountain** in Scotland, to the nearest 10 metres?

5   What was the **life expectancy** for women born in **1970**?

6   There are **129 Members of the Scottish Parliament**. How many are **women**?

7   Work out the difference between the **highest** and **lowest temperatures** ever recorded in Scotland.

8   In **2001**, what **percentage** of the Scottish population lived in the **Central Belt** (which is made up of the **Industrial West** and the **Forth Estuary**)? Give your answer to the **nearest 5%**.

9   Work out the **area** of Scotland that is taken up by **National Parks**.

10 How many people **living in Scotland** are aged **under 16**?

CfE Outcomes

Through practical activities which include the use of technology, I have developed my understanding of the link between compass points and angles and can describe, follow and record directions, routes and journeys using appropriate vocabulary.
MTH 2-17c

I can display data in a clear way using a suitable scale, by choosing appropriately from an extended range of tables, charts, diagrams and graphs, making effective use of technology.
MTH 2-21a/MTH 3-21a

I can find the probability of a simple event happening and explain why the consequences of the event, as well as its probability, should be considered when making choices.
MNU 3-22a

Links with
Technology, Social Studies

Task 2

The five highest mountains in Scotland are

- Ben Nevis
- Ben Macdui
- Braeriach
- Cairn Toul
- Sgor an Lochan Uaine

1 Complete the tally chart for the letters **A, E, I, B, C, N** and **others** to count the number of times the letters are used in the mountain names.

Letter	Tally	Frequency
A		
E		
I		
B		
C		
N		
Others		

2 Using centimetre-squared paper, draw a **bar chart** to represent the information from question 1.

3 Now use your results to answer these questions.

   a **One letter** is chosen at **random** from the mountain names. What is the **probability** that it is the letter **R**?

   b **One letter** is chosen at **random** from the mountain names. What is the **probability** that it is **not** the letter R?

Task 3

Here is a **map** of Scotland.

Complete the sentences as **accurately as possible**, using one of the following **directions**.

> north   east   south   west   north-east   north-west   south-east   south-west

1   Edinburgh is _____ of Glasgow

2   Perth is _____ of Stirling

3   Dundee is _____ of Aberdeen

4   Inverness is _____ of Glasgow

5   Dundee is _____ of Wick

6   Stranraer is _____ of Dunfermline

Bricklaying patterns

This is a typical house brick.

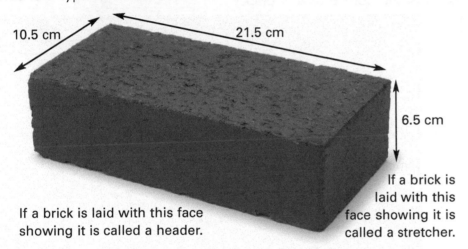

10.5 cm

21.5 cm

6.5 cm

If a brick is laid with this face showing it is called a header.

If a brick is laid with this face showing it is called a stretcher.

When bricks are used to build walls they need to be laid so that they are strong enough to support floors and roofs. Bricks laid in an interesting pattern can also add to the visual attractiveness of a property.

There are many different arrangements of bricks. For example, one of the most common is the arrangement below, left, which is called Stretcher bond. The arrangement below, right, is called Flemish bond and is made with alternating stretchers (the bricks that lie along the wall) and headers (the bricks that lie across the wall). It is a very strong arrangement and is used when a wall is two bricks thick.

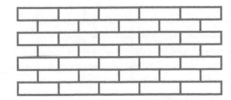
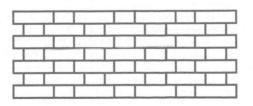

Imagine that these brick patterns continue in all directions. What are the symmetries?

The Stretcher bond has the following symmetries:

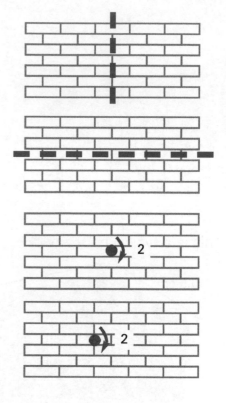

- a vertical line of symmetry through the centre of any brick

- a horizontal line of symmetry through the centre of any brick

- rotational symmetry of order 2 through the centre of any brick or the centre of any vertical join.

Task 1

Describe the **symmetries** to be found in the **Flemish bond** pattern.

Task 2

Describe the **symmetries** to be found in **each** of the following brick patterns.

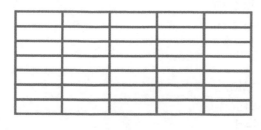

Stack bond

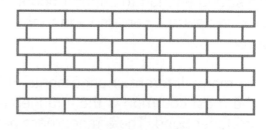

English bond

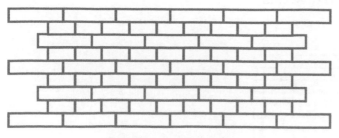

English cross bond

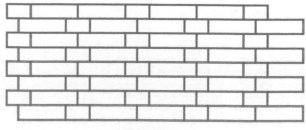

Monk bond

Task 3

Bricks are also used in **pavements** and are laid in various patterns.

Four **paving patterns** are shown below. Describe the **symmetries** of each one.

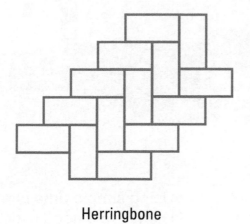

Herringbone

Basketweave

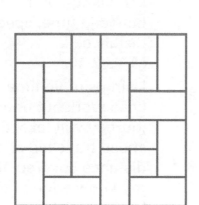

Pinwheel

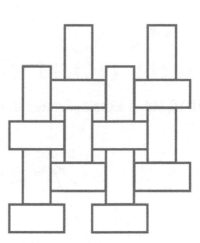

De laRobia weave

Deliveries

Courier companies deliver parcels for their customers. They usually have several large warehouses throughout the country. The parcels are sent to these warehouses, then sent out to smaller depots for local delivery. It is important for the warehouse manager to plan routes that are as short as possible, but that cover all the depots.

Warm-up questions

On-time couriers (OTC Ltd) makes deliveries from its warehouse to three local depots.

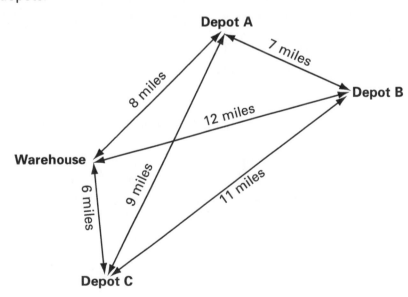

1   How far does a van travel if it goes from the **warehouse** to **depot A** and then to **depot C** and back to the **warehouse**?

2   If the van travels at **30 mph** how long does it take to travel **from the warehouse to depot B**?

CfE Outcomes

Using simple time periods, I can give a good estimate of how long a journey should take, based on my knowledge of the link between time, speed and distance.
MNU 2-10c

Using simple time periods, I can work out how long a journey will take, the speed travelled at or distance covered, using my knowledge of the link between time, speed and distance.
MNU 3-10a

3   If a van can carry **35 parcels**, how many journeys would be needed to take
200 parcels?

4   How much does it cost for a van to travel **150 miles** at **80p** per mile?

5   How much does it cost for a van to travel **150 miles** every day for **1 working week
(6 days)**?

Task 1

As warehouse manager for OTC Ltd, refer back to the plan and work out the **shortest route**
for a driver who must visit **all three depots** and **return to the warehouse**.

Task 2

The company opens a new depot called **depot D**.

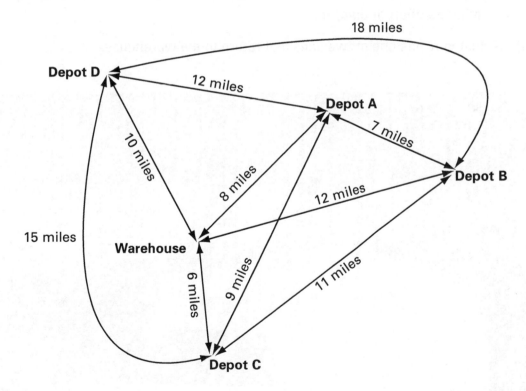

Work out the **shortest route** for a driver who must visit **all four depots** and **return to the
warehouse**.

Task 3

The company decides to use **two delivery vans**.

Work out the best way that they could **share deliveries** to **all four depots** and **return to the warehouse**.

The **running costs per mile** for each van is **80p**.

Compare the **daily costs** for using **one and two vans**.

State any **advantages** or **disadvantages** of using one or two vans.
State clearly any **assumptions** you make.

Task 4

A delivery company has its warehouse in **Glenrothes**.

Deliveries are made to depots in **Paisley**, **Hawick**, **Stranraer** and **Inverness**.

Use the **internet** to work out a **mileage chart** or **diagram**.

Decide the best routes, if the company has one or two vans that return to the warehouse.

Water

Each person in Scotland uses an average of about 150 litres (33 gallons) of water per day.

The pie chart gives a breakdown of how that water is used.

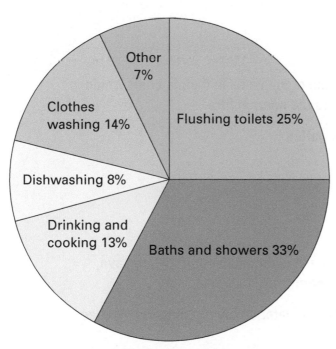

Pie chart sections:
- Other 7%
- Flushing toilets 25%
- Clothes washing 14%
- Dishwashing 8%
- Drinking and cooking 13%
- Baths and showers 33%

All households pay for the water they use each year. The amount paid is based on the council tax band for each household. The table shows the water charges for the different council tax bands.

Council tax band	Water supply	Waste water collection	Combined services
A	£121.44	£140.94	£262.38
B	£141.68	£164.43	£306.11
C	£161.92	£187.92	£349.84
D	£182.16	£211.41	£393.57
E	£222.64	£258.39	£481.03
F	£263.12	£305.37	£568.49
G	£303.60	£352.35	£655.95
H	£364.32	£422.82	£787.14

Task 1

1   Work out how many **litres** of water **per day** each of the items in the pie chart would be for an **average person**. Give your answers to the **nearest litre**.

2   Approximately how many **litres** are there in a **gallon**?

Task 2

A typical **house brick** is **21.5 cm** by **10.25 cm** by **6.5 cm**. It is recommended that, to save water, households place a house brick **in the toilet cistern**.

A litre is **1000 cm³**. A toilet cistern holds about **6 litres** of water. An average person flushes a toilet about **6 times** a day.

Work out approximately how much water is saved per person **in a year** by adding a house brick to the toilet cistern.

Task 3

1 A house is in **Council tax band B**. What are the charges for their **water supply** and their **waste water collection**?

2 A house is in **Council tax band G**. What are the charges for their **water supply** and their **waste water collection**?

3 How much **more in total** does a householder in a house in **Council tax band F** pay compared with a householder in a house in **Council tax band C**?

Task 4

Some homes have water meters fitted to measure the amount of water used. The meter is provided free by Scottish Water, but the householder must pay for the installation, which may be between £230 and £1000. The water used is charged at about 0.5p per litre (including waste water charges), and there is a also a fixed charge which is related to the size of the water supply.

1 If 3 people living in a house all use the average amount of water each day, work out how much they would pay each year if the house is fitted with a water meter, and if the water is charged at 0.5p per litre.

2 If the house is in Council tax band F, what would be the difference in the cost if the house did not have a water meter?

Task 5

This is part of a newspaper article.

1 Approximately how many **litres** of water are used in a **bathtub**?

2 A person drinks **4 cups** of coffee a **day**. Approximately how many **bathtubs** of water each **year** will be used to produce the ingredients for the coffee this person drinks?

Shocking but true! If you are a stickler for environmental issues, even you might be surprised at the amount of water you use! A recent British report has suggested that the average person in the UK 'uses' 4500 litres of water daily – that is roughly 58 bathtubs full of water every day.

The report estimated that while each person in the UK uses around 150 litres of mains water every day, they consume about 30 times as much in 'virtual' water embedded in food, clothes and other items.

For example, just one tomato from Morocco takes 13 litres of water to grow while the ingredients in a cup of coffee collectively use 140 litres. A shirt made from cotton grown in Pakistan or Uzbekistan soaks up 2700 litres of water.

Task 6 (extension)

Produce a poster or a slide show to show water usage and how we can all conserve water by following a few simple rules.

Pilots of aircraft have to operate according to air law, in the same way that drivers of vehicles have to obey the *Highway Code*.

Flight levels

Aircraft flying quite high use flight levels (FLs) to describe the altitude of the aircraft.

When air traffic controllers and pilots transmit or write down flight levels they leave out the last two zeros of the altimeter reading.

> **FL 50** (flight level five zero) means an altitude of **5000 feet**.
> **FL 45** (flight level four five) means an altitude of **4500 feet**.
> **FL 150** (flight level one five zero) means an altitude of **15 000 feet**.
> Flight levels are only used for **intervals of 500 feet** up to **FL 250** (25 000 feet) and then only **every 1000 feet** above **FL 250**.
> For example, FL 35, FL40, FL45, up to FL 250, then FL 260, FL270 and so on.

Safety in the UK

The semi-circular rule

For safety reasons, aircraft flying in different directions need to fly at different heights.

Aircraft flying in controlled airspace in the UK must fly at a cruising flight level according to the semi-circular rule, as shown in this table and in the diagram opposite.

CfE Outcomes

Having investigated navigation in the world, I can apply my understanding of bearings and scale to interpret maps and plans and create accurate plans, and scale drawings of routes and journeys.
MTH 3-17b

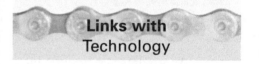

Links with Technology

(Magnetic track) 000° to 179°	(Magnetic track) 180° to 359°
FL 10	FL 20
FL 30	FL 40
FL 50	FL 60
FL 70	FL 80
FL 90	FL 100
FL 110	FL 120
FL 130	FL 140
FL 150	FL 160
Up to FL 410, then every 4000 feet	Up to FL 400, then every 4000 feet

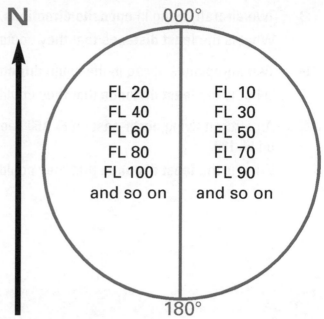

Questions

Use the information given above to answer these questions.

1 What is **6500 feet**, written as a **flight level**?

2 What is **FL 280** written in **feet**?

3 Explain why **FL 34** is not used.

4 An aircraft is flying on a **bearing of 070°**.
Using the **semi-circular rule**, at which of these **flight levels** could it fly?
FL 100   FL 110   FL 120   FL 130

5 An aircraft is flying on a **bearing of 290°**.
Using the **semi-circular rule**, at which of these **flight levels** could it fly?
FL 100   FL 110   FL 120   FL 130

6 An aircraft is flying **westwards**.
Using the **semi-circular rule**, at which of these **flight levels** could it fly?
FL 100   FL 110   FL 120   FL 130

7 An aircraft is flying **eastwards**.
Using the **semi-circular rule**, at which of these **flight levels** could it fly?
FL 100   FL 110   FL 120   FL 130

8 Two aircraft, flying in **opposite directions**, are using the **semi-circular rule**.
What is the **least distance** that they could be **apart**?

9 Two aeroplanes, flying in the **same direction**, are using the **semi-circular rule**.
What is the **least distance** that they could be **apart**?

10 An aircraft flying **north-east** on **FL 150** sees another aircraft flying **south-east** on **FL 190**.
What is the **least distance** that they could be **apart**?

Darts

Darts is a popular game. Players take it in turns to throw three darts at a board like the one shown here.

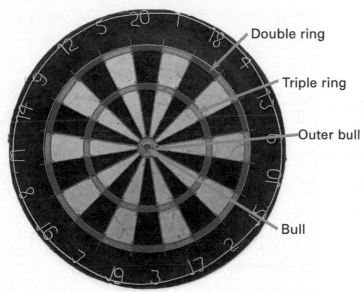

Double ring

Triple ring

Outer bull

Bull

CfE Outcomes

Having determined which calculations are needed, I can solve problems involving whole numbers using a range of methods, sharing my approaches and solutions with others.
MNU 2-03a

I can use a variety of methods to solve number problems in familiar contexts, clearly communicating my processes and solutions.
MNU 3-03a

Depending where the dart lands, it can score any number, from the lowest score of 1 to the highest possible score of 60.

There are 20 sectors. Each sector is worth the score shown on the outside.

Around the outside of each sector is the double ring. If a dart lands in this ring it scores double the value of the sector. Halfway in there is another, smaller ring called the triple ring. A dart landing in this ring scores treble the value of the sector.

In the very centre is a small circle surrounded by a small ring. The middle circle is called the bull and scores 50. The small ring is called the outer bull and scores 25.

The most common darts game is 501. Players take it in turn to throw their darts and their total score is deducted from 501. The first player to finish, by reducing the score to 0, wins. However, to finish, the last dart to reduce the score to 0 (or to achieve the total of 501) must be a double. The bull counts as a double.

If the final three darts give a total that is greater than the number needed to finish, which means the total scored would be over 501, these throws are void.

Professional darts players are experts at working out finishes for various scores. For example:

- 164 would require treble 20, treble 18 and the bull
- 105 could be treble 15, single 20 and double 20 or treble 19, single 18 and double 15.

Some scores, such as 164, only have one three-dart finish, whereas others, such as 105, have many different three-dart finishes.

Questions

In these questions, **T** means a **treble score**, **D** means a **double score** and **S** means a **single score**.

1 What are the following **totals** for **three darts**?

   a T18, D6, S3   b T17, D16, S8
   c D17, S17, S3   d T15, D10, D6
   e T19, S18, D4

2 A score of **100** has many **three-dart finishes**. Find a **three-dart finish for 100** in which the **final dart** is:

   a D20   b bull   c D16.

3 a What is the **largest** number that **can** be scored with **three darts**?
   b What is the **smallest** number that **cannot** be scored with **one dart**?

4 All the numbers up to **98** can be scored with a **two-dart finish**. Show a **two-dart finish** for **each** of these numbers.

   a 75   b 98   c 3

5 Show that **99** is the smallest score that needs a **three-dart** rather than a two-dart **finish**.

6 Assuming the dart hits the board, the **smallest number** that can be scored with **one dart** is **1** and the **largest** is **60** (T20). Work out which scores **between 1 and 60 cannot** be scored with **one dart**.

7   Look at the dartboard on the right.

The numbers **20 and 3** are on **opposite sectors**. Adding these gives a total of 23.

Which numbers on **opposite sectors**, when added, give a **total that is**:

a   a total of 24 (there are 2 answers)

b   a square number

c   the smallest total?

8   Look at the dartboard again.

The numbers **20 and 3** are on **opposite sectors**. Subtracting the **smaller** from the **bigger** gives a **difference of 17**.

Which numbers on **opposite sectors** have a **difference that is**:

a   12 (there are 2 answers)

b   a square number

c   the biggest possible?

9   Look at the dartboard again.

The numbers **20** and **5** are on **adjacent sectors**. Adding these gives a **total of 25**.

Which numbers on **adjacent sectors**, when added, give a **total that is**:

a   25 (there are 2 more answers)

b   a square number other than 25

c   19 (there are 4 answers)

d   the largest possible?

10   The **total** of **all the numbers** on the dartboard is:

$1 + 2 + 3 + 4 + 5 + 6 + 7 + 8 + 9 + 10 + 11 + 12 + 13 + 14 + 15 + 16 + 17 + 18 + 19 + 20$

The answer is **210**. Explain an **easy way** to work this out **without adding up all the numbers**.

Hint: $10 + 11 = 21$ and $9 + 12 = 21$.

Extension question

The lowest possible number of darts that can be thrown to score **501** and **finish on a double** is **9**.

Work out **how this can be done**. There is more than one answer.

Bridges

In this activity, you will find out about the Forth Road Bridge, then you will design and build a bridge of your own.

You will need the **Data sheet: Bridges** (p 120).

Task 1

Use the **Data sheet: Bridges** to answer these questions.

1   How **long did it take** to **build** the Forth Road Bridge?

2   In 1978, how many **vehicles** crossed the bridge each month, on average? Give your answer to the **nearest 5000**.

3   How much **concrete** was used to build the Forth Road Bridge?

4   Which **three bridges** in the table are within **200 metres** of the **length** of the Forth Road Bridge?

5   In high winds, the Forth Road Bridge is sometimes closed to high-sided vehicles. How much further is it to travel from **Edinburgh** to **Kirkcaldy** via the **Kincardine Bridge** than using the Forth Road Bridge? Give your answer in kilometres. (A distance of 5 miles is approximately 8 kilometres.)

6   For how many years was the Forth Road Bridge the **longest single-span suspension** bridge in Europe?

7   How much **longer** is the span of the Forth Road Bridge than the **Tancarville Bridge** in France?

8   Calculate the **total tolls** charged in **1997** in a one-minute period if the traffic consisted of **7 motor bikes, 16 cars, 3 lorries and 1 bus**.

9   Calculate the **average toll** paid by vehicles crossing the **northbound** side of the bridge in 2000. What does this tell you about the **different types** of vehicle crossing the bridge?

10   What **percentage** of the bridges listed in the table are in the **UK**?

Task 2

Here are some facts about a different bridge. Use the information given in the table on the data sheet (p 120) to work out which of the ten bridges this is about. Explain how you worked out your answer.

> This bridge is 40% longer than the Forth Road Bridge and was the longest single-span suspension bridge in the world for 17 years. The bridge was officially opened by the same member of the Royal Family who opened the Forth Road Bridge.

Task 3

Design and build a **bridge** that has a **span** of **at least 30 centimetres**.

Draw up a **design specification** before you start to construct the bridge.

Your bridge should be **strong enough** to support the **weight of a calculator** at any point. Your bridge must be **at least 10 cm** above the ground at **all points between the supports**.

Your design specification should include any **measurements** that you make.

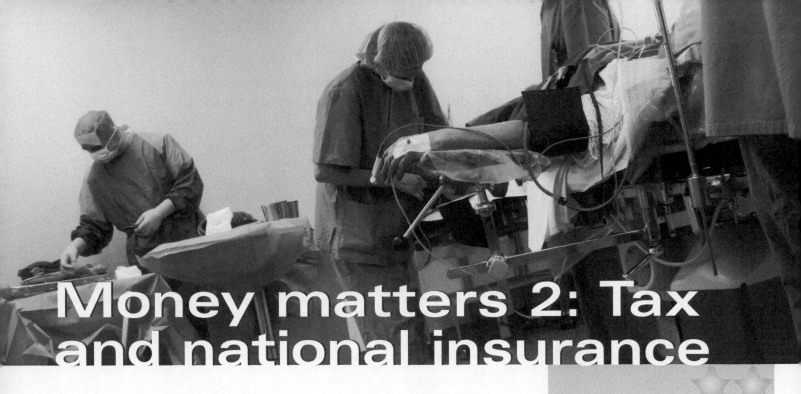

Money matters 2: Tax and national insurance

All workers who earn money have to pay income tax. Most people pay tax by the Pay as you earn (PAYE) system. This means tax is taken on wages each week or month, so that the worker does not get a huge tax bill once a year. Taxes are used to fund things such as the National Health Service (NHS), the armed forces, the police and government expenses. In addition, workers also pay national insurance contributions (NIC). These are used to cover pensions, sick pay and benefits.

The amount a worker earns, before any deductions are made, is called the gross income. The amount a worker receives is called the net income or take-home pay.

Each year the Government publishes a budget that sets all the tax rates and allowances.

Personal allowance

Everyone is allowed a personal tax allowance. This is the amount anyone is allowed to earn before they start paying income tax. The difference between gross income and personal tax allowance is the taxable income.

This is the tax allowance table for 2011–12.

Income tax allowance	2011–12
Personal allowance	£7475

Tax rates

Tax is charged at different rates, depending on how much a worker earns overall. There is a basic tax rate of 20%, a higher tax rate of 40% on taxable earnings over £35 000, and an additional rate of 50% on taxable earnings over £150 000.

These rates apply to taxable income, not gross income.

This table shows the tax rate for 2011–12.

Tax rates on *taxable* income	2011–12
Basic rate: 20%	£0 – £35 000
Higher rate: 40%	£35 000 – £150 000
Additional rate: 50%	Over £150 000

National insurance (NI)

The rate paid in National Insurance varies, depending on total earnings. For everything between the primary threshold and the upper earnings limit, a worker will pay 12% in National Insurance. For example, someone earning £400 per week will pay 12% of (£400 – £139), i.e. 12% × £261 = £31.32. A worker earning over the upper earnings limit per week will pay an additional 2% for everything over £817. For example, someone earning £900 per week will pay 12% of (£817 – £139) plus 2% of (£900 – £817), i.e. 12% × £678 + 2% × £83 = £83.02.

	2011–12
Upper earnings limit	£817
Primary threshold	£139
Rate between primary threshold and upper earnings limit	12%
Rate above upper earnings limit	2%

This table shows the NIC rate for 2011–12 for earnings per week.

Example 1

Mr T earns £380 per week. Work out his net income.

Answer

Gross income = £380 × 52 = £19 760 per annum

Taxable income = £19 760 – £7475 = £12 285

Tax paid = 20% of £12 285 = £2457

Weekly tax paid = £2457 ÷ 52 = £47.25

NI contribution = 12% of (£380 – £139) = £28.92

Net weekly income = £380 – (£47.25 + £28.92) = £303.83

Example 2

Ms C earns £82 000 per annum. Work out her monthly take-home salary.

Answer

Monthly salary = £82 000 ÷ 12 = £6833.33

Taxable income = £82 000 − £7475 = £74 525

Tax paid = 20% of £35 000 + 40% of (£74 525 − £35 000) = 0.2 × £35 000 + 0.4 × £39 525 = £22 810

Monthly tax payable = £22 810 ÷ 12 = £1900.83

Weekly earnings = £82 000 ÷ 52 = £1576.92

NI contribution = 12% of (£817 − £139) + 2% of (£1576.92 − £817) = £96.56

Monthly NI contribution = £96.56 × 52 ÷ 12 = £418.43

Net monthly income = £6833.33 − £1900.83 − £418.43 = £4514.07

Questions

1   Miss P earns **£220 per week**. Work out her **weekly take-home pay**.

2   Mr Q earns **£44 000 per annum**. Work out his **monthly net salary**.

3   Mrs R earns **£2500 per month**. Work out her **monthly net salary**.

4   Ms S earns **£550 per week**. Work out her **weekly net wage**.

5   Mr T earns **£380 per week**. His wife, Mrs T, earns **£130 per week**.
    Work out their **joint weekly take-home pay**.

6   Mr U earns **£52 000 a year**. His partner, Ms V, earns **£58 000 a year**. What is their **joint monthly net income**?

7   Mr W earns **£3000 a month**. Six months into the year he gets a pay rise of **£500 a month**. Work out by how much **extra** his monthly take-home salary increases after the rise.

8   Mr X earns **£2000 per month**. He pays taxes and national insurance at the **normal rates** for **8 months**. He is then made redundant and doesn't get another job for the rest of the year.

   a   Work out how much **tax he would pay** on an annual income of **£24 000**.

   b   Work out how much **tax he would pay** on an annual income of **£16 000**.

   c   Work out how much **tax rebate he should get** at the **end of the year**.

Revision planning

As you approach exams, you may decide to plan for revision so that you can do as well as possible. In this activity, you will make **revision plans** for two students and for yourself.

This table shows the subjects that the students are taking.

Core subjects
English
Mathematics
Physics
Chemistry

Other subjects
Art and Design
French
Geography
History
Computing and Information Science
Modern Studies
Music
Physical Education

Task 1

Suppose you are a student in **S4** and you are **planning revision** for your examinations.

You need to plan a **7-day revision programme**. Include **all of the subjects** listed on page 41.

Each day you have up to **six half-hour revision slots** to fill.

You have already decided that **core subjects** will have **2 hours per week** and **other subjects** will have **1 hour per week**.

For any **core subject**, you should revise for no **more than 1 hour per day** (two slots together or two separate slots).

For **other subjects**, you should revise for **no more than $\frac{1}{2}$ hour per day**.

You **should not revise** any **core subject** immediately **after another core subject**. (Ignore breaks.)

This plan will leave you **7 hours free** for **other activities**.

You could use a chart like this to create your timetable.

	6.00 pm–6.30 pm	6.30 pm–7.00 pm	7.00 pm–7.15 pm	7.15 pm–7.45 pm	7.45 pm–8.15 pm	8.15 pm–8.30 pm	8.30 pm–9.00 pm	9.00 pm–9.30 pm
Monday								
Tuesday								
Wednesday								
Thursday			Break			Break		
Friday								
Saturday								
Sunday								

Task 2

Now suppose you decide to do even more revision. You will still use up to six half-hour slots each **weekday** (Monday to Friday) but you decide that on **Saturday and Sunday** you will use **up to six 45-minute slots** with **30-minute breaks**.

Your **weekend** revision will start at **10.00 am each day**.

You have up to **six half-hour revision slots** to fill on each **weekday** (Monday to Friday) and up to **six 45-minute slots** to fill on **Saturday and Sunday**.

Copy the tables opposite. Complete the timings in the table for **Saturday and Sunday**, starting from **10.00 am** each day. Now fill in your timetable using the following rules.

You have decided that:

- **core subjects** will have **3 hours per week**
- **French** will have **2 hours per week**
- **history** will have **1.5 hours per week**
- **other subjects** will have **1 hour per week**.

You should revise **any core subject** for **no more than:**

- **1 hour per day** (two slots together or two separate slots) on **weekdays**
- **1.5 hours per day** (two slots together or two separate slots) on **Saturday or Sunday**.

You should revise **other subjects** for **no more than:**

- $\frac{1}{2}$ **hour per day** on **weekdays**
- **45 minutes** on **Saturdays or Sundays**.

You **should not revise** any **core subject** immediately **after another core subject**. (Ignore breaks.)

This will leave you **5.5 hours free** for **other activities**.

	6.00 pm–6.30 pm	6.30 pm–7.00 pm	7.00 pm–7.15 pm	7.15 pm–7.45 pm	7.45 pm–8.15 pm	8.15 pm–8.30 pm	8.30 pm–9.00 pm	9.00 pm–9.30 pm
Monday			Break			Break		
Tuesday								
Wednesday								
Thursday								
Friday								

	10.00 am–							
Saturday								
Sunday								

Task 3

Using your own subjects, make up your own revision rules and plan your own revision timetable.

At the gym

A gym opens from 6 am to 8 pm each day from Monday to Friday and from 9 am to 5 pm on Saturday and Sunday.

In the men's changing rooms at the gym there is a shower room with five showers.

The diagram shows the plan of the shower room.

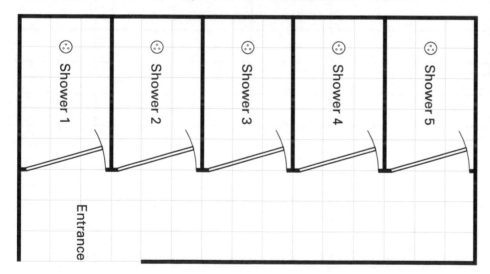

Shower 1 Shower 2 Shower 3 Shower 4 Shower 5

Entrance

Each shower is equipped with a soap dispenser that holds 500 ml of soap when full.

The dispensers are filled every morning, before the gym opens. At 9 am one of the staff checks the dispensers.

This diagram shows the level in each dispenser at 9 am one Monday.

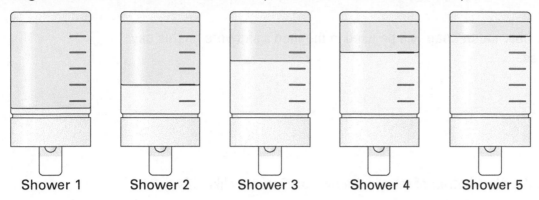

Shower 1   Shower 2   Shower 3   Shower 4   Shower 5

This is the distribution of men entering the gym for that day.

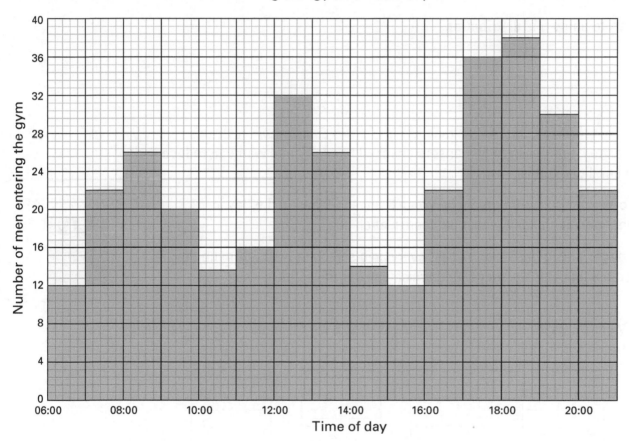

Task 1

Give an explanation for the **levels of soap** in **each dispenser** at **9 am**.

Task 2

Give an explanation for the **distribution of men entering the gym** during the day.

Task 3

Work out approximately **how much soap** will be used in the men's showers on this day. Give your answer in **litres**.

Task 4

The manager describes the **distribution** of men in the gym on Saturday like this.

'There are about 20 members each hour entering the gym, with a slight increase over lunchtime and in the early afternoon.'

Draw a **bar chart** to show this **distribution**.

Task 5

The manager has to have the soap dispensers checked regularly and wants to **reduce the time** her staff spend doing this.

Please use shower

a   If all showers were used **equally** between 6 am and 9 am on the day in question, how much soap would be used in each one?

b   The manager puts in a sensor so that every time someone walks through the entrance a sign lights up suggesting which shower they should use.

The following Monday, when the showers are checked at 9 am these are the levels of soap in the dispensers.

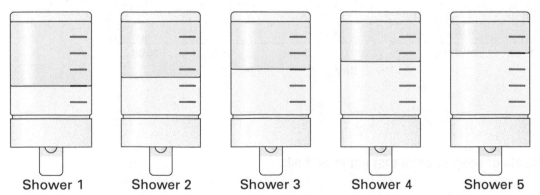

| Shower 1 | Shower 2 | Shower 3 | Shower 4 | Shower 5 |

Given that the **distribution** of men entering the gym **is the same as for the previous Monday**, work out the **latest time** the manager should get the dispensers checked so that **none of them runs out**.

46

Task 6 (extension)

To stay healthy an average man needs 2500 calories and an average woman needs 2000 calories a day.

Moderate cycling on the exercise bike, moderate rowing on the rowing machine, aerobics, using the treadmill at walking pace and light weightlifting will burn about 7 calories per minute.

Fast cycling on the exercise bike, heavy rowing on the rowing machine, kick boxing, using the treadmill at jogging pace and heavy weightlifting will burn about 10 calories per minute.

Design an exercise program that last one hour and burns approximately 500 calories.

Use the internet to find out how many calories various sportsmen use.

For example

- Riders in the Tour de France each day
- Marathon runners during a marathon
- Professional footballers during a match

Money matters 3: Loans and APR

Almost always, when someone takes out a loan, they have to **pay interest**.

Companies that offer credit and loans must always state the interest rate they charge, and this is usually given as the **annual percentage rate (APR)**.

The APR shows what your interest payment would be if the interest was compounded and paid annually instead of monthly (or any other period). The APR is used so that customers can **compare** different credit and loan deals.

When the APR is calculated it takes into account the interest the customer has to pay, the timing and amount of the payments, any fees for setting up the deal, and any fees for payment protection if the lender makes this compulsory.

The law says that all lenders must tell you what their APR is before you sign an agreement.

Generally, the lower the APR the better the deal for the customer.

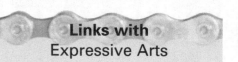

Look at these advertisements for loans.

TopSERVICE PERSONAL LOANS

THE TYPICAL RATE IS 7.8% APR ON ALL LOANS FROM £5000 TO £25 000

APPLY ONLINE FOR A TS LOAN.

MONTHLY REPAYMENTS.

Gordon's cheap loans

Wipe out debt. Get a loan. Call in today.

Borrow from £5000 to £100 000

Consolidate all your debts into 1 lower monthly payment

Raise money for home improvements, a holiday, etc

Borrow up to 80% of your property value.

APR from as low as 8.5%

The Instant Cash Loans Company

One-off payment option.

Borrow cash now up to £1000.

Pay back on pay-day.

One simple phone call away.

You receive	Our interest	Repayable
£80.00	£20.00	£100.00
£160.00	£40.00	£200.00
£240.00	£60.00	£300.00
£320.00	£80.00	£400.00
£400.00	£100.00	£500.00

APR 1284% based on 31 days

The Lending Bank

You can borrow up to £20 000.
You can choose from fixed monthly repayments of one to five years.
Enjoy the peace of mind of our optional loan repayment insurance.

APR 9.9%
Apply online or call at your local branch.

££££££££££

Cash Around the Corner Loans

Apply now for cash loans from £50 to £500 with no hidden charges

Your local loan shop, established 100 years

Example:
Loan amount: £100, 57 weekly repayments of £3 Total amount payable £171
189.2% APR

Example 1

Calculate the APR on **a loan of £1000** for **one year** at **10% interest**, repaid at the **end of the year**.

Answer

Interest = 10% of £1000 = £100

Total repayment at the end of the year = £1100

The APR is 10%.

Example 2

Calculate the APR on a **loan of £1000** for **one year** at **10% interest, repaid every month** in **equal monthly instalments**.

10% interest = 10% of £1000 = £100

Total repayment made throughout the year = £1100

Because the money is paid back throughout the year, not all of the £1100 is available for the whole year, so the APR is much higher. It is about 20%.

Warm-up questions

1   Look at the advertisement for **TopService Loans**.
   a   What is the **least** you can **borrow**?
   b   What is the **most** you can **borrow**?
   c   How much is **7.8%** of **£1000**?
   d   How much is **7.8%** of **£5000**?

2   Look at the advertisement for **Gordon's Cheap Loans**.
   a   How **often** are the **repayments**?
   b   If you own a property worth **£100 000**, what is the **most** that can you borrow?
   c   How much is **8.5%** of **£10 000**?
   d   How much is **8.5%** of **£50 000**?

3   Look at the advertisement for **The Instant Cash Loans Company**.
   a   How much is the **APR** (based on **31 days**)?
   b   What is **£20** as a **percentage** of **£80**?
   c   Why is the **APR** so **high**?

4   Look at the advertisement for **Cash Around the Corner Loans**.
   a   How **often** are the **repayments**?
   b   What is the **most** you can borrow?
   c   Why is this **APR** so **high**?

5 Look at the advertisement for **The Lending Bank**.

   a What is **most** that you can borrow?

   b What is the **longest time** you can take to **pay off the loan**?

   c **What else** do they offer to go with the loan?

Task 1

Copy and complete the table. The first company has been done for you.

Name of company	Online, branch or shop	Amount that can be borrowed	Repayments: one payment, weekly or monthly	APR
TopService Personal Loans	Online	£5000 to £25 000	Monthly	7.8%
Gordon's Cheap Loans				
The Instant Cash Loans Company				
Cash Around the Corner Loans				
The Lending Bank				

Task 2

Imagine that you wanted to take out a **loan**.

Prepare a report **analysing** the advertisements shown.

For each advertisement, state any **good points** or any **bad points**.

Explain which are **easy** to follow.

Design your own advertisement, using the **best features** of the advertisements shown.

Stickers

Lots of companies sell packs of stickers: for example, of football clubs or the Harry Potter books and films. Collectors buy them in packs and stick them into albums until they have collected a **full set**. The packs are put together **randomly** so nobody knows in advance what stickers will be in the pack. Most people end up with several stickers the same, just missing a couple of stickers to complete the set. There are internet sites where stickers can be bought or swapped.

In this activity you will build a **mathematical model** to study the process of completing a set of stickers.

CfE Outcomes

I can use a variety of methods to solve number problems in familiar contexts, clearly communicating my processes and solutions.
MNU 3-03a

Links with
Technology

Task 1

This task will model how long it takes to get a full set of stickers when there are a **total of five** to collect and they are sold in **packs of three**.

Work in pairs or small groups.

Take a sheet of A4 card and divide it into **20 equal rectangles**, five rows and four columns.

Now decide what your sticker set is going to be about.

You will need to have **five different things** or **people** in your set. For example, you can choose five of your favourite pop artists or just five colours.

The cards shown in the example are marked H, M, B, L and Mg. Can you guess what these represent?

Mark **each** thing or person on **four** of the cards.

Now cut the cards out carefully, so all of them are **exactly the same shape and size**.

Turn the cards over and **shuffle them** or lay them out on the table and mix them up, so you can't see what is on the front.

Now choose **three cards at random** and write down what you get.

Put the cards back, face down, **shuffle them again** and take **three more cards at random**. Again, note the cards you get.

Keep on doing this until you have picked one of each type.

Each time you pick a **set of three cards** it represents buying a **pack of stickers**.

For example:

Pack 1	H, L, L	So far you have collected H and L.
Pack 2	M, L, H	So far you have collected M, H and L.
Pack 3	Mg, H, M	So far you have collected M, H, L and Mg.
Pack 4	M, H, H	So far you have collected M, H, L and Mg.
Pack 5	M, B, H	Now you have collected the whole set M, H, L, Mg and B.

This has modelled the situation where it took **five packs** to collect a **whole set**.

Repeat this task **five times** in total.

What was the **smallest number** of packs it took to get a whole set?

What was the **greatest number** of packs it took to get a whole set?

What was the **average number** of packs it took to get a whole set?

Your teacher may collect in the results from **all groups** and work out an **average** for the **whole class**.

Task 2

It is generally claimed, but denied by the companies that make stickers, that they do not print exactly the same number of each sticker in the set.

This task will model how long it takes to get a full set of stickers when there are a total of **five to collect** and they are sold in **packs of three** but where the **numbers of each type of sticker printed are not equal**.

Proceed as before but, this time, only mark **two cards with H**, mark **three with M**, mark **four with B**, mark **five with L** and mark **six with Mg**.

Repeat the experiment, again taking **three cards at a time** until you have a **full set**.

Repeat this task **five times in total**.

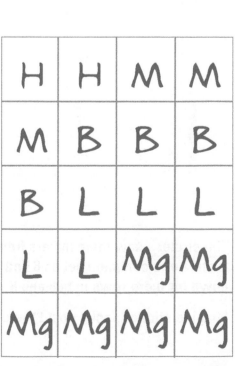

What was the **smallest number** of packs it took to get a whole set?

What was the **greatest number** of packs it took to get a whole set?

What was the **average number** of packs it took to get a whole set?

You should have found it took **a lot longer** to get a full set.

Task 3 (extension)

Now make up your own mathematical model for the **total number of stickers in a set** and the **number of stickers in a pack**, using a computer **spreadsheet** program.

Imagine there are **20 stickers in total in the set** and they are sold in **packs of four**.

Open up a blank spreadsheet and in cell A1 type in **=INT((RAND() * 20) + 1)**, then press **return**. Copy this formula down **column A** for about **20 rows**. Now highlight all the cells with numbers in **column A** and copy this across for **four columns**.

This should look something like the screen below (without the circles).

	A	B	C	D
1	5	10	4	13
2	9	15	4	7
3	16	2	14	19
4	2	1	16	5
5	16	20	13	19
6	16	2	19	20
7	20	17	4	3
8	7	14	1	7
9	9	20	10	4
10	4	16	20	1
11	12	15	18	2
12	2	19	16	20
13	12	10	14	3
14	7	3	3	11
15	5	12	1	9
16	2	5	1	15
17	7	4	7	14
18	7	17	15	10
19	11	4	5	14
20	7	17	12	14

> This model was for **20 stickers in packs of four**. If it had been for **25 stickers in packs of five**, the spreadsheet formula would be **=INT((RAND() * 25) + 1)** and the formula would be copied across for **five rows**.

The circles show the numbers **from 1 to 20**. So **even after 20 packs** have been bought there is still not a complete set as **6 and 8 are missing**. The formula would need to be **copied down for more rows** until 6 and 8 are obtained.

Imagine **sticker packs cost £1 each**. How much will it **cost to complete a set**?

Money matters 4: Savings and AER

The **annual equivalent rate** (**AER**) is the interest rate quoted on **interest paid** on savings and investments.

It shows what the interest return would be if the interest was compounded and paid annually instead of monthly (or any other period).

The AER is used so that customers can **compare** different savings accounts.

Gross AER is the rate of interest **before deduction** of income tax.

Net AER is the amount of interest payable **after deduction** of 20% tax for basic-rate taxpayers.

Here are some advertisements for savings accounts.

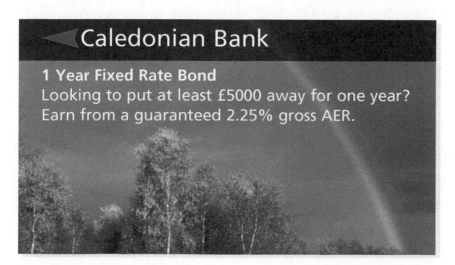

Caledonian Bank

1 Year Fixed Rate Bond
Looking to put at least £5000 away for one year?
Earn from a guaranteed 2.25% gross AER.

Castle
BUILDING SOCIETY

eSaver Direct

Committed online savers will enjoy a great variable rate of 2.30% gross/AER if you make just one withdrawal in the first year or 2.50% gross/AER if you make none. This includes a bonus of 0.75% gross/AER for 12 months.

A lower interest rate of 0.10% gross/AER applies for each calendar month in which a withdrawal takes place.

DUNBLANE BUILDING SOCIETY

The minimum balance for this account is £500.

A fixed interest rate of 1.50% Gross/AER is payable.

No penalty for withdrawals.

Saltire BANK

£1000+ invested. 3.00% gross pa/AER (variable) including 1.40% bonus (gross) for 12 months.

Bonus of 1.40% (gross) paid for first 12 months provided you maintain a balance of at least £1000.

Portree Building Society

1, 2 or 3 Year Fixed Term Bond

Annual:		
3.00% Gross pa	2.40% Net pa	3.00% AER
Monthly:		
2.96% Gross pa	2.37% Net pa	3.00% AER

Interest rates are fixed until:
1-Year Option Fixed Term
2-Year Option Fixed Term
3-Year Option Fixed Term

Interest applies to balances of £1 plus.

A **withdrawal penalty** equivalent to 180 days' loss of interest on the amount withdrawn applies for the fixed-rate period.

Warm-up questions

1   Look at the advertisement for the **Caledonian Bank**.
   a   From the information given, what is a **fixed rate bond**?
   b   Is it possible to invest **£2000** in this bond?
   c   How much is **2.25%** of **£1000**?
   d   How much is **2.25%** of **£5000**?

2   Look at the advertisement for the **Castle Building Society**.

   a   Where do you **invest** in this account?

   b   What is the **interest rate** without **withdrawals** (AER)?

   c   What is the **interest rate** in a month when a **withdrawal** takes place (AER)?

3   Look at the advertisement for the **Dunblane Building Society**.

   a   How much do you need to **invest** in this account?

   b   What is the **interest rate** (AER)?

   c   What is the **penalty** for **withdrawals**?

4   Look at the advertisement for the **Saltire Bank**.

   a   What is the **least amount** that you can **invest** in this account?

   b   When is the **bonus paid**?

5   Look at the advertisement for the **Portree Building Society**.

   a   What is the **penalty** for **withdrawals**?

   b   How much do you need to **invest** to receive **interest**?

   c   What does **Net pa** mean?

Task 1

Some accounts have **restrictions** on whether you can invest in them.

For each account, if possible, work out the **annual interest** for an investment of

   a   £500

   b   £1000

   c   £5000

Where it is not possible to invest the sum of money, explain the reason why.

Assume that no withdrawals are made.

Task 2

Imagine that you are the marketing manager of a different building society or bank.

Prepare a **report**, analysing the advertisements shown.

For each advertisement, state any **good points** or any **bad points**.

Explain which advertisements are **easy to follow**.

Design **your own advertisement**, using the **best features** of those shown.

Shuffleboard

Shuffleboard is a game in which players take turns to propel red or yellow discs along the playing area. Each player shoots **four discs**. Points are scored when the discs land in the areas marked with numbers in the scoring triangle.

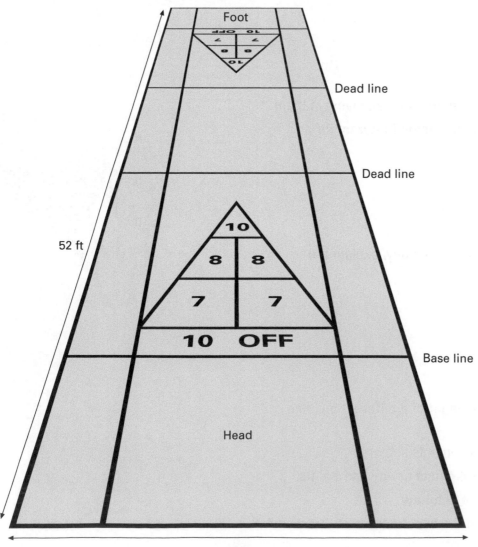

This is a plan view of the scoring triangle. It is drawn to scale with 1 square representing 1 foot.

Possible scores are **10**, **8**, **7** and **10 off**, which means a disc landing in this area causes 10 points to be **deducted** from the player's total.

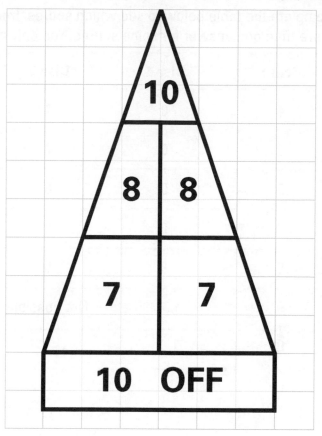

Task 1

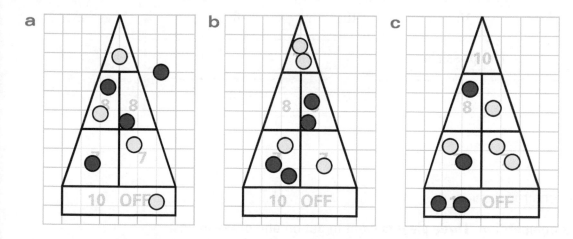

Work out the **scores** for the **red** and **yellow discs** in these diagrams.

Task 2

A score of 4 is the **lowest possible positive score**. This can be obtained by scoring 0, 7, 7 and **−10**. Zero (0) means the **disc does not score**.

Scores of **5, 6, 7** and **8** are also **possible** but a score of **9** is **impossible**.

The **maximum possible score** would be **40**, with four scores of 10.

Complete the table below to see which scores, **from 4 to 40**, are possible. There may be more than one answer for some scores. You only need to find one.

Score	Disc 1	Disc 2	Disc 3	Disc 4
4	0	7	7	−10
5	0	7	8	−10
6	0	8	8	−10
7	0	0	0	7
8	0	0	0	8
9		Not possible		
10				
11				

Score	Disc 1	Disc 2	Disc 3	Disc 4
38	10	10	10	8
39		Not possible		
40	10	10	10	10

Task 3

Shuffleboard is very popular on cruise ships.

Two teams decide to have a **competition**. **Three players** want to get on one of the teams. They each record their scores for at least their last **30** games.

Alf records his scores in a **stem-and-leaf diagram**.

Key: 1 | 2 represents a score of 12

```
0 | 7   8
1 | 4   5   5   6   6   7   7   8   8
2 | 0   1   1   2   2   3   4   5   7   8   8   8
3 | 0   0   1   1   2   3   4
```

Brenda records her scores in a **grouped table**.

Score	Frequency
$16 \leqslant$ score $\leqslant 20$	9
$21 \leqslant$ score $\leqslant 25$	18
$26 \leqslant$ score $\leqslant 30$	3

Clara records her scores as a **frequency diagram** like this.

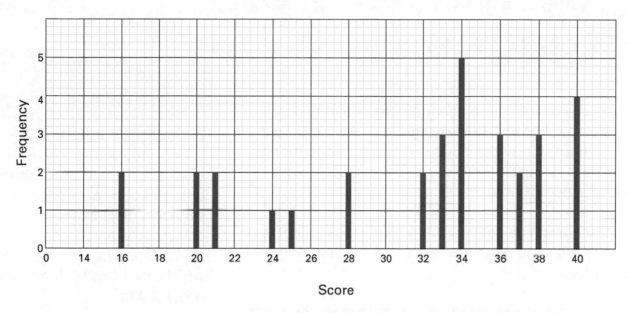

You are the team captain. Explain fully why you would pick **Clara** for the team. Who would you pick as the reserve, and why? (Hint: calculate and compare the range, mean, mode and lowest scores).

Task 4 (extension)

The scoring triangle consists of an **isosceles triangle**, **four trapezia** and a **rectangle**.

Work out the **area** of **each of these shapes**.

Assuming that where the disc lands is random, calculate the **probability** of each score by working out the **area of each score** as a **decimal fraction** of the **total area**.

Money matters 5: Mortgages

Buying a house is probably the biggest purchase that you will make. You will probably need a **mortgage** to make the purchase. A mortgage is a **loan** to buy the house. You borrow the money, usually from a bank or building society, and pay it back, with **interest**, over a period of time.

Mortgages are **secured** against your home. This means that if you cannot afford the repayments the lender could sell your home to get their money back.

Although there are many different types of mortgage, the most common is the **repayment** mortgage.

Repayment mortgage

The amount you pay each month is made up of capital (some of the amount you borrowed) and interest.

The amount is calculated so that you are guaranteed to have repaid everything you owe by the end of the borrowing period.

> ## Example 1
> You **borrow £100 000** at **5% interest** for **20 years**.
> The **monthly repayment** = £659.96
> The **interest paid** = £58 390.40
>
> ## Example 2
> You **borrow £100 000** at **4% interest** for **25 years**.
> The **monthly repayment** = £527.84
> The **interest paid** = £58 352.00

Example 3

You **borrow £100 000** at **6% interest** for **10 years**.
The **monthly repayment** = £1110.21
The **interest paid** = £ 33 225.20

Variable rate and fixed rate mortgages

Mortgages may also be **variable** rate or **fixed-rate**.

If you have a variable rate mortgage, the lender can change the interest rate, which means the cost of your mortgage can go **up or down**.

If you have a fixed rate mortgage, the lender fixes the interest rate for an agreed period of time, usually for the first one, two or three years, and then changes the arrangement, either to a variable rate or to offer a new fixed rated.

This table shows the advantages and disadvantages of a fixed rate mortgage.

Advantages	Disadvantages
You know exactly how much you will have to pay for the whole period of time.	You will probably pay more than the variable rate on offer at the beginning.
You are not affected by interest rate rises.	You are not affected by interest rate falls.

Task 1

The table shows the **monthly repayments** for an **interest rate of 5%**.

	£110 000	£120 000	£130 000	£140 000	£150 000
15 years	£883.14	£963.42	£1043.71	£1123.99	£1204.28
20 years	£735.56	£802.43	£869.29	£936.16	£1003.03
25 years	£650.40	£709.52	£768.65	£827.78	£886.91

1   How much is the **monthly repayment** for a **£120 000** mortgage taken out for **20 years**?

2   How much is the **monthly repayment** for a **£150 000** mortgage taken out for **15 years**?

3   How much would be the **total repaid** for a **£110 000** mortgage taken out for **25 years**?

4   How much would be the **total repaid** for a **£140 000** mortgage taken out for **15 years**?

5 How much **more** does it cost per **month** for a **£120 000** mortgage than a **£150 000** mortgage taken out for **20 years**?

This table shows the **monthly repayments** for an **interest rate of 7%**.

	£110 000	£120 000	£130 000	£140 000	£150 000
15 years	£1006.45	£1097.95	£1189.44	£1280.94	£1372.43
20 years	£865.27	£943.93	£1022.59	£1101.25	£1179.91
25 years	£786.60	£858.11	£929.61	£1001.12	£1072.63

6 How much **more** is the **monthly repayment** for a **£130 000** mortgage taken out for **20 years** at **7%** than at **5%**?

7 How much **more** is the **monthly repayment** for a **£140 000** mortgage taken out for **25 years** at **7%** than at **5%**?

8 How much **more** would be the **total repaid** for a **£140 000** mortgage taken out for **15 years** at **7%** than **5%**?

9 Suppose you want a mortgage for **£300 000**.
Use the table to work out the **difference** in the **monthly repayments** for a **15-year** and a **20-year** mortgage with an interest rate of **7%**.

10 Suppose you want a mortgage for **£270 000**.
Use the table to work out the **monthly repayments** for a **25-year** mortgage.

Task 2

Do an internet search on **mortgage calculator**. Many different companies provide mortgage calculators but they all work in a similar way. When you find a mortgage calculator put in some details. For example:

Enter mortgage details		Monthly repayment	
Mortgage amount (£)	100000	583.33	Interest only
Years left on mortgage	25	706.78	Repayment
Annual rate of interest (%)	7		

Calculate   Clear

You can also use the **second part** of the calculator to find out what will happen if the **rate of interest increases**. You would need to know if you could still afford the mortgage.

What if the interest rate changes?

Would you still be able to pay? Select an interest rate to find out.

Choose rate increase		**Monthly repayment**	
Rate increase	3%	883.33	Interest only
		908.70	Repayment

Recalculate   Clear

Find a calculator that will compare mortgages when the **interest rate changes** as above.

Compare the **monthly repayments** for **different mortgages**.

Find out how **monthly repayments** for **different mortgages** change when the **interest rate changes**.

Prepare a **report** to present to the class. Make sure that you **comment** on the effect of:

- taking out a mortgage **for a longer period**
- **borrowing more**
- **increases** in the **interest rate**, for example, comment on the effect of a **1% increase** on a mortgage of **£100 000** taken over **25 years**.

Task 3

Advertisements for mortgages have to include the following warning statement.

> **Your home may be repossessed if you do not keep up repayments.**

Design your own **advertisement** for a **mortgage**. Give as much **information** as you can making all the facts as **clear** as possible.

Make sure that you:

- tell them **who you are**
- give **contact details** (make these up)
- state whether it is **fixed rate** or **variable rate**
  - If **fixed rate** state what happens to the interest rate **at the end of the fixed rate period**.
  - Does it revert to **variable rate**, is a **new fixed rate offered** or are **both options** available?
- state the **interest rate**
- state the **period** of the mortgage available (this could be a range, for example, 15 to 25 years).

Remember to include the **warning statement**.

Time zones

As you know from your science lessons, the Earth rotates on its axis. Therefore, sunrise occurs at different times in different parts of the world.

Because every part of the world considers that the day begins when they see the Sun rise, the Earth is divided into 24 time zones. This is why when it is 12 noon (midday) in London it is only 7 am in New York but 3 pm in Moscow. Businesses that trade with foreign countries need to be aware of the time differences so they can contact people when they are also at work. Travellers need to know about time differences so that they can plan journeys.

Time zones are based on **Coordinated Universal Time (UTC)**. Previously, time zones were based on **Greenwich Mean Time (GMT)** but UTC has been used since 1961. In Britain, for all everyday purposes, time is based on GMT from the last Sunday in October to the last Sunday in March. For the rest of the year **British Summer Time** is used, which is equivalent to **GMT + 1** or **UTC + 1.** Note that GMT and UTC do not vary throughout the year.

Times in most places around the world are whole numbers of hours ahead or behind UTC but in some places the difference may be in half or quarter hours.

The standard way of showing a time zone is UTC ± n (or GMT ± n), where n is the difference in hours. For example, the local time at various places at 12:00 midday UTC would be:

- Los Angeles, California, United States: UTC − 8 = 04:00
- New York City, United States: UTC − 5 = 07:00
- London, United Kingdom, in December: UTC = 12:00
- Paris, France: UTC + 1 = 13:00
- Moscow, Russia: UTC + 3 = 15:00
- Karachi, Pakistan: UTC + 5 = 17:00

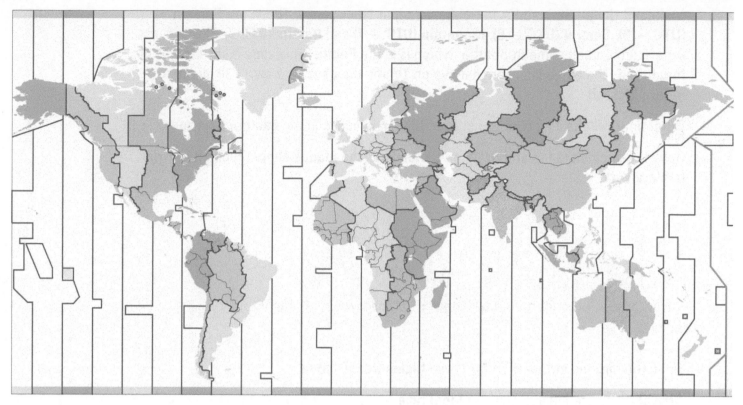

This map shows the time zones around the world.

Questions

1   A flight from **London to New York** takes **6 hours and 30 minutes**. It takes off from London at **8:30 am**. What will be the **local time** in **New York** when it lands?

2   A woman is in **Moscow** on business. She rings her family in the **United Kingdom** at **7 pm local time** in **Moscow**. What time does her family get the call?

3   A woman travels from **Los Angeles** to **Paris**. She leaves home in **Los Angeles** at **midday local time** and arrives at her sister's house in **Paris** at **5:30 pm local time** the **following day**. How long did her journey take, door to door?

4   A business has offices in New York, London and Paris.

        The **New York** office is open from **8 am to 6 pm** local time.

        The **London** office is open from **9 am to 5 pm** local time.

        The **Paris** office is open from **10 am to 7 pm** local time.

For how many hours in the day are **all three offices open** at the **same time**?

5   The time in **Melbourne**, Australia is **UTC + 11**. At **9 am local time** Mr Dundee, who lives in **London**, rings his daughter in Melbourne to wish her a happy Christmas. What time of the day does **his daughter** get the call?

6 Mainland United States of America covers four times zones. These are **Eastern (UTC − 5)**, **Central (UTC − 6)**, **Mountain (UTC − 7)** and **Pacific (UTC − 8)**. An American football game in Seattle, which is in the **Pacific time zone** kicks off at **2 pm local time**. The game is broadcast live on TV and the coverage starts **30 minutes before kick off**.

What time will the TV coverage start in **Miami**, which is in the **Eastern time zone**?

7 Copy these clock faces and draw on the **times in the places shown** when it is **1 pm in New York**.

| New York | London | Los Angeles | Moscow | Karachi |

8 Fill in the missing information for these **flights from London**.

Arrival times are always given in **local time**.

Flight	Destination	Departure	Length of flight	Arrival Time
AU1	Moscow	06:45	3 h 10 m	
AU2	Paris	10:25	1 h 05 m	
AU3	New York	11:10		14:30
AU4	Los Angeles	08:50		10:20
AU5	Karachi		8 h 15 m	20:30

9 A family arranges a three-way video call on New Year's Eve. Laura is in **Melbourne (UTC + 11)**, Andrew is in **Newcastle (UTC + 0)** and Carol is in **Miami (UTC − 5)**. Laura makes the call **five minutes before midnight, local time**. What are the **local times** for Andrew and Carol?

10 A family in Vancouver, **Canada, (UTC − 8)** listens to a BBC radio programme **live** on the internet. The program is broadcast in the **UK** from **8 pm to 10 pm**. Between what **local times** will the family listen to the programme in **Vancouver**?

Planning a bedroom

Imagine you are moving to a new house, where you will need to plan how you want your bedroom.

For each task below, the scale drawing represents a bedroom. The plan shows the position of the door.

You need to decide how to fit out the bedroom with new furniture. First, you will need to decide what size of bed to put in the room. You will also need to include sufficient wardrobes and drawers to make the room practical to use.

Use centimetre-squared graph paper to draw up your own scale drawing for the plan for each bedroom. Choose a scale so that your diagram takes up the whole sheet of graph paper.

Write a brief report explaining any decisions you take. For example, say why you chose the position of the furniture and other items and explain the number of drawers you wanted.

The **Data sheet: Planning a bedroom** (p 121) gives details of furniture sizes.

Task 1

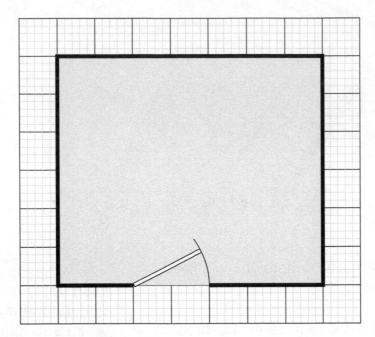

Scale: 1 centimetre represents 0.5 metres

Task 2

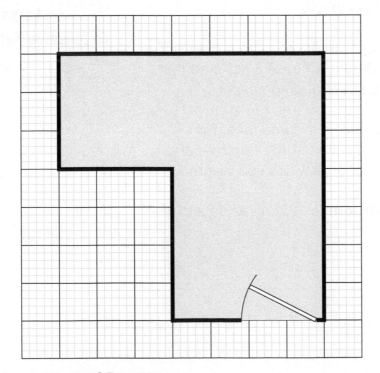

Scale: 1 centimetre represents 0.5 metres

Task 3

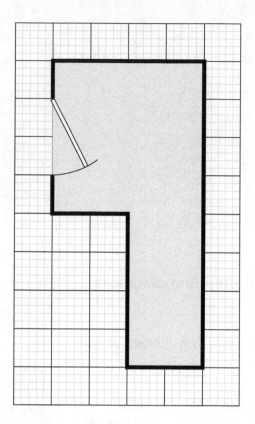

Scale: 1 centimetre represents 0.5 metres

Stopping distances

For this activity you will need **Data sheet: Stopping distances** (p 122).

The *Highway Code* includes a table of **safe stopping distances** and gives advice to drivers about stopping a car at different speeds.

The **stopping distance** consists of two phases. The first is the **thinking distance**. This is the distance the vehicle travels while the driver's brain reacts to the situation and realises that the brake needs to be applied. The second phase is the **braking distance**, which is the distance the vehicle travels before it comes to a stop.

Data sheet: Stopping distances shows the stopping distances for various speeds.

Task 1

1   How many **kilometres** are equivalent to **10 miles**?

2   What is the **thinking distance** for a speed of **50 mph**?

3   What is the **braking distance** for a speed of **96 km/h**?

4   What is the **total stopping distance** for a speed of **40 mph**?

5   What is the **total stopping distance** for a speed of **112 km/h**?

6   Estimate the **thinking distance** for a speed of **45 mph**.

7   Estimate the **total stopping distance** for a speed of **65 mph**.

8   How many **car lengths** should be left between cars travelling at **70 mph** to be totally safe?

9   The graph shows the **thinking distance** (blue), **braking distance** (red) and **total stopping distance** (black) for cars travelling at **20 mph** and **30 mph**. Copy and complete the graph.

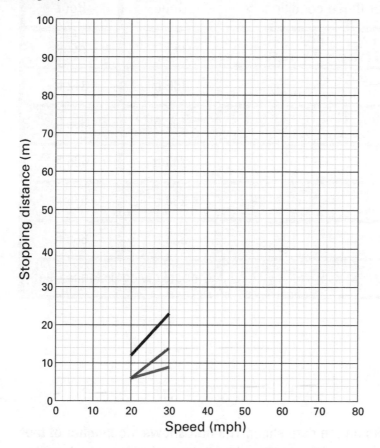

10   The **maximum speed limit** in Britain is **70 mph**. Extend the graph to estimate the **thinking**, **braking** and **total stopping distances** for **80 mph**.

Task 2

The table on the next page, from an American website, recommends leaving a **3-second gap** between cars in **good driving conditions** and a **6-second gap** in **poor driving conditions**.

1   Work out the **middle distances** for each pair, to get the distance travelled at **30 mph**, **40 mph**, **50 mph**, **60 mph** and **70 mph** for both a **3-second** and **6-second** gap.

2   There are **3.28 feet** in **1 metre**. Convert the distances in question 1 to **metres**. Give your answers to the **nearest metre**.

3   Compare **these distances** to those based on the **UK stopping distances**.

4   Another website claims a formula for calculating the stopping distance is:

$$D = \frac{x^2}{20} + x$$

where **D** is the **stopping distance** in **feet** and **x** is the **speed** in mph.
Compare the distances from the **formula** with those shown on the **data sheet**.

Three-second rule		Safe interval should be >	3 seconds	6 seconds
Speed	Distance travelled	For these conditions >	Good	Poor
25 mph	37 ft per second		111 ft	222 ft
35 mph	52 ft per second		166 ft	312 ft
45 mph	66 ft per second		198 ft	396 ft
55 mph	81 ft per second		243 ft	486 ft
65 mph	96 ft per second		288 ft	576 ft
75 mph	75 ft per second		333 ft	666 ft
			Safe following distance (feet)	

Task 3

Look at the **emergency braking graphs** on the data sheet. The first shows the **number of feet** travelled under **emergency braking** for speeds from **60 mph** and the **probability of a fatality**. This means that in a crash at **60 mph** there is a **50%** chance that the person involved will **die**. The second shows the **time** it takes to **brake from 60 mph** and the **probability of a fatality**.

These questions relate to **time** and **distance** after **emergency braking** from **60 mph**.

1   How **far** has a car travelled **before** the speed has **reduced** to **20 mph**?

2   How **long** does it take for the **speed** to **reduce** to **20 mph**?

3   a   At what **speed** does the **chance** of a **fatality** become almost **zero**?
    b   How **far** has the car **travelled** when it **reaches this speed**?
    c   How **long** does it take to **reach this speed**?

4   How **long after braking** does the **chance** of a **fatality** reduce to **0.25**?

5   What is the **speed** after **1.5 seconds**?

6   How **far** has the car travelled during this **1.5 seconds**?

7   How **long** does it take the car to travel **20 feet**?

8   How **far** does the car travel in **1 second**?

9   What is the **speed** when the **chance** of a **fatality** is reduced to **1 in 10**?

10 Complete the graph for **time** against **stopping distance** for **emergency braking** from **60 mph**. Two points have already been plotted.

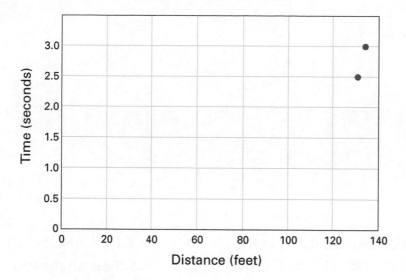

Emergency braking from 60 mph

Task 4 (extension)

Write a **letter** to the prime minister saying why you think the **speed limit** outside schools during school hours should be reduced to **20 mph**.

Climate change

These graphs show levels of carbon dioxide (CO_2) in the atmosphere and global temperature change over the last 150 years.

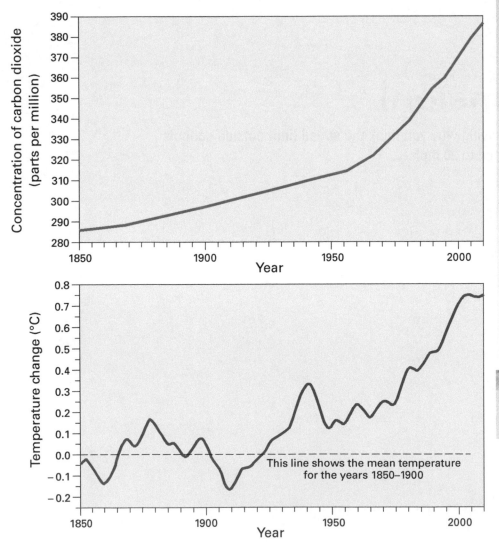

CfE Outcomes

When analysing information or collecting data of my own, I can use my understanding of how bias may arise and how sample size can affect precision, to ensure that the data allows for fair conclusions to be drawn.
MTH 3-20b

I can evaluate and interpret raw and graphical data using a variety of methods, comment on relationships I observe within the data and communicate my findings to others.
MTH 4-20a

Links with
Technology, Science, Social Studies

These graphs show levels of carbon dioxide (CO_2) in the atmosphere and global temperature change over the last 400 000 years.

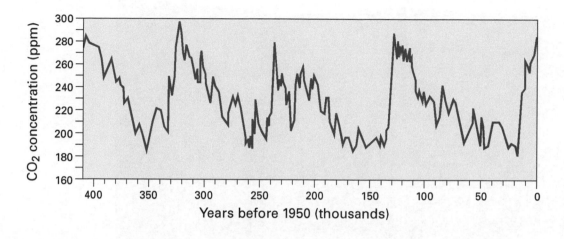

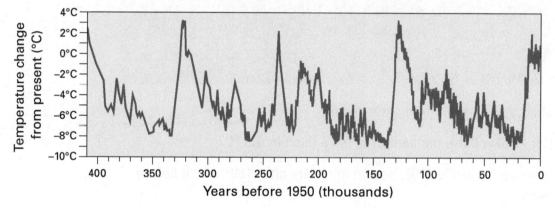

Task 1

Use both pairs of graphs to show that there is a link between the amount of carbon dioxide in the atmosphere and the global temperature.

Task 2

Here are some predictions and facts about the environment.

- **Temperatures could rise** by about **2.5 °C** by the year **2100**.
- **Sea levels could rise** on average **0.5 cm per year** for the **next 100 years**.
- The **five warmest years** on record are, in **descending order of warmth**, 2010, 2005, 1998, 2002, 2003.
- **Glaciers are melting**. Between 1961 and 2007, the world's glaciers lost **2200 cubic miles** of ice.
- The **USA** has **4.5%** of the **world's population** but produces **25%** of the **carbon dioxide pollution** from **fossil-fuel burning**.
- The **UK** produces 434 million tonnes of waste each year.
- The **UK** produces 5.2 million tonnes of hazardous waste each year.
- Householders in the **UK** produce almost **30 million tonnes** of waste on average **each year**. Of this waste, **73% goes to landfill**, even though **90% of it is recoverable** and could be **recycled**.

Answer the following questions, using the information given.

1. How many years **after 2010** will the temperatures have risen by 2.5 °C?

2. The **third warmest** year on record is **1998**. How many years after 1998 was it before this **record was broken**?

3. The **UK** produces **434 million tonnes of waste** each year. Write **434 million** in figures.

4. By how many **metres** will **sea levels rise** in the **next 100 years**?

5. At the end of October 2011, the world population was approximately 7 billion. Use this to estimate the **population of the USA**. Give your answer to the **nearest 100 million**.

6. How many **tonnes** of **UK householders' waste** go to **landfill**?

7. What **fraction** of **UK householders' waste** that goes to landfill is **recoverable** and could be **recycled**?

8. There are about **9 million cubic miles** of ice on Earth. What **percentage** of this was lost **between 1961 and 1997**?

Task 3

Use the **graphs** to support the argument that **global warming is not happening**.

Use **other information** to support the argument that **global warming is caused by human behaviour**.

Growing, growing, grown...

This activity is about how some mathematical series grow very quickly. This is known as **exponential growth**.

Before you start, find a piece of scrap paper. It is said that, no matter how thin the paper may be, it is impossible to fold it more than seven times. Try it and see. If you have a tissue, try it with that. Did you manage more than seven folds?

The reason it gets so difficult to fold is that the number of layers you are folding doubles every time. When you start you have one layer. This is folded in half to give two layers. These are then folded to give four layers and so on. The number of layers each time is given by:

Fold	0	1	2	3	4	5	6	7	8	...
Layers after fold	1	2	4	8	16	32	64	128	256	...

So as you make the seventh fold you are trying to fold 64 layers. Imagine trying to fold 64 pieces of A4 paper in half! It can't be done.

Now think about this problem, which is very similar to the folding activity.

You have a piece of thin paper that is infinitely big. You cut it in half and place the two sheets together. You then cut these two sheets in half and place them together to give four sheets. You keep on doing this until you have made 50 cuts. How high will the resulting pile of paper be?

★ ★ ★

CfE Outcomes

Having explored number sequences, I can establish the set of numbers generated by a given rule and determine a rule for a given sequence, expressing it using appropriate notation.
MTH 3-13a

Having explored how real-life situations can be modelled by number patterns, I can establish a number sequence to represent a physical or pictorial pattern, determine a general formula to describe the sequence, then use it to make evaluations and solve related problems.
MTH 4-13a

Links with
Technology

You will have met some simple powers such as **square** and **cube**. Powers are used to write long, repetitive multiplication calculations in a shortened way.

For example, $5 \times 5 \times 5 \times 5 \times 5 \times 5 \times 5 \times 5 \times 5$ can be written as **5^9**

and   **6^4** means $6 \times 6 \times 6 \times 6$.

The series for the number of sheets above can be written as:

$$2^0 \quad 2^1 \quad 2^2 \quad 2^3 \quad 2^4 \quad 2^5 \quad 2^6$$

and so on. Note that in mathematics anything raised to the power 0 is always 1 and anything raised to the power 1 is the same as itself. So:

$$2^0 = 1, \; 2^1 = 2, \; 2^2 = 2 \times 2 = 4, \; 2^3 = 2 \times 2 \times 2 = 8$$

and so on.

Questions

1   Continue these power series up to 10 terms.

   a 3^0 3^1 3^2 3^3 3^4 …   …   …   …   …

          1   3   9   …   …   …   …   …   …   …

   b 5^0 5^1 5^2 5^3 5^4 …   …   …   …   …

          1   5   25   …   …   …   …   …   …   …

   c 3×2^0 3×2^1 3×2^2 3×2^3 3×2^4 …   …   …   …   …

          3   6   12   …   …   …   …   …   …   …

2   Complete the following table, which shows the numerical value of powers of 10.

Number	0.0001		0.01	0.1	1		100		
Fraction			$\frac{1}{100}$	$\frac{1}{10}$	1	$\frac{10}{1}$			
Power		10^{-3}		10^{-1}	10^0	10^1	10^2	10^3	10^4

3   **64 balls** are dropped into the **maze** shown in the diagram.
      At each point marked with a letter **half the balls** go to the left and **half** go to the **right**.
      So, at A, **32 balls** go to the **left** and 32 go to the **right**.
      At B, **16 balls** go to the **left** and 16 go to the **right**.
      Each ball lands in one of **five trays** at the bottom of the maze.

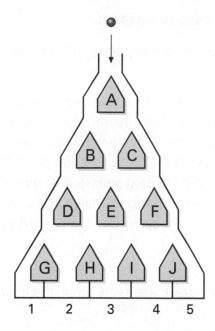

a   Explain why **16** balls end up in **tray 2**. You may find it easier to copy the picture and mark how the 64 balls **split up** as they go through the maze.

b   Work out how many balls end up in **each** of the **other trays**.

4   Imagine an alien from outer space arrives in Britain with a happiness bug on **1 January**. The number of people infected **doubles** each day. So on 2 January **four** people are infected, on 3 January **eight** people are infected and so on.

The population of Britain is approximately **62 million** people.

a   On approximately what day of the year is the **whole population** infected? **Hint:** Use trial and improvement.

b   Another alien also lands on **1 January** and spreads a love bug in the same way. The world has a population of **7 billion**. Will the **whole world** be in love before St Valentine's Day (**14 February**)?

5   How many **squares** are there on a **chessboard**?

The answer is not 64.

Start by counting **how many squares** there are in one **small square**.   □

Silly question, the answer is **one**.

Now count how many squares there are in a **2 x 2** board.   ⊞

Not such a silly question, there are **five squares**:
four small 1 x 1 squares and one larger 2 x 2 square.

Now explain why there are **14 squares** in a **3 x 3 board**.

Count how many squares there are in a **4 x 4 board** and then a **5 x 5 board**.

Put your results in a **table** and see if you can see a **pattern**.

Once you see the pattern you can work out how many squares there are in an **8 x 8 board**.

6   The man who invented the game of **chess** took it to his king and taught him the game. The king was very pleased and said to the man: 'As a reward I shall give you your **weight in gold**.' The man said: 'I do not deserve such a prize, I shall be happy to have enough food to last my family the rest of their days. I ask that you place **1 grain of rice in first square** of the chessboard, **2 grains in next square**, **4 grains in the next square** and **so on until the 64th square** is filled.' The king smiled and said: 'You are wise enough to invent that game but you are not so wise in the reward you demand. Your wish is granted.'

Who got the **better deal**?

You will need the following information (or look these facts up on the internet to get the current values).

- One **kilogram** of **rice** contains **60 000 grains** of rice.
- A **tonne** is 1000 **kilograms**.
- A **tonne** of **rice** costs **£400**.
- Gold is worth **£30** a **gram**.
- The **average** weight of a man is **80 kg**.

Task 1 (extension)

Investigate the **Fibonacci** series:

0, 1, 1, 2, 3, 5, 8, 13, 21, 34, …

- How many **terms** will there be **before 1000** is reached?
- What happens when you keep on **dividing consecutive pairs** of numbers:

   $1 \div 1 = 1, 2 \div 1 = 2, 3 \div 2 = 1.5, 4 \div 3 = 1.33$

   and so on? What number do you eventually **end up with**?
- Where does the Fibonacci series occur in **nature**?

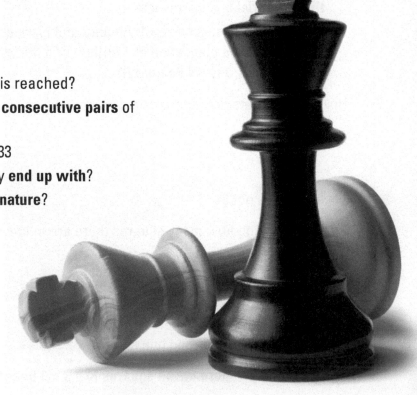

Chocolate cake

Chocolate cake is a very popular dessert. There are at least two different ways of preparing it.

The table below shows the guideline daily amounts (GDAs) of foods for women, men and children aged 5–10 years.

	Women	Men	Children 5–10 years
Energy (kcalories)	2000	2500	1800
Fat (g)	70	95	70
Of which saturated fat (g)	20	30	20
Carbohydrate (g)	230	300	220
Of which total sugars (g)	90	120	85
Protein (g)	45	55	24
Dietary fibre (g)	24	24	15
Salt (g)	6	6	4

Here are two recipes for chocolate cake, with additional information.

Chocolate cake recipe 1 (24 servings)	Chocolate cake recipe 2 (12 servings)
Ingredients	**Ingredients**
120 ml milk	285 g plain flour
85 g plain chocolate (grated)	5 g baking powder
300 g caster sugar	3 g salt
2 eggs	120 ml water
170 g sour cream	84 g dark chocolate
5 g baking powder	115 g butter
15 ml water	300 g caster sugar
285 g plain flour	2 eggs
115 g butter, softened	5 ml vanilla extract
	155 g sour cream
Nutrition (per serving)	**Nutrition (per serving)**
Total fat: 7.1 g	Total fat: 15.3 g
Cholesterol: 31 mg	Cholesterol: 62 mg
Salt: 230 mg	Salt: 747.5 mg
Total carbohydrate: 24.4 g	Total carbohydrate: 46.3 g
Dietary fibre: 0.5 g	Dietary fibre: 1.5 g
Protein: 2.2 g	Protein: 4.2 g

Task

You are preparing a study session with **five** of your classmates and you want to bake them a cake, and then have some left over for your **family**. Compare the two **recipes**, the **cost of ingredients** and the **guideline daily amounts**.

Decide which **recipe** to use. Write a **report** which **justifies** your decision. Include in your report any **advantages** and **disadvantages** of using each recipe.

Cost of ingredients for chocolate cake	
Milk chocolate	74p for 150 g
Dark chocolate	74p for 150 g
Caster sugar	£1.28 for 1 kg
Eggs	£1.46 for 6
Sour cream	£1.13 for 300 ml
Baking powder	72p for 170 g
Plain flour	98p for 1.5 kg
Butter	£1.60 for 250 g
Milk	88p per litre
Vanilla extract	74p for 30 ml

Venting gas appliances

For this activity you will need the **Data sheet: Venting gas appliances** (pp 123–124).

Gas appliances must have adequate venting systems to dispose of the waste gases, including carbon monoxide (CO), safely into the atmosphere. When gas appliances do not have adequate venting systems, people can die of carbon monoxide poisoning.

Governments set down minimum standards for venting systems. These must be followed by builders and gas fitters so that gas appliances are safe.

The data sheet shows the minimum requirements for a gas vent pipe that emerges through a roof. It also gives a table of data for the maximum permitted output in thousands of **British thermal units (Btus)** per hour.

> A **British thermal unit** (Btu) is a measure of **energy**.
>
> **Inches** are abbreviated as, for example, **9"**.
>
> **Feet** are abbreviated as, for example, **6'**.

Look at the first page of the data sheet. The diagram shows a typical configuration for a gas vent pipe emerging from a roof.

It must be at least 8 feet from a vertical wall.

The **pitch** of a roof is a measure of the **steepness**. It is calculated by measuring how many inches a roof rises vertically for every 12 inches horizontally.

CfE Outcomes

I can apply my understanding of scale when enlarging or reducing pictures and shapes, using different methods, including technology.
MTH 3-17c

Having recognised similarities between new problems and problems I have solved before, I can carry out the necessary calculations to solve problems set in unfamiliar contexts.
MNU 4-03a

I can apply my knowledge and understanding of measure to everyday problems and tasks and appreciate the practical importance of accuracy when making calculations.
MNU 4-11a

Links with
Science, Technology

Task 1

Refer to the **Data sheet: Venting gas appliances**. Use the table showing the **roof pitch and minimum height of vent** to answer these questions.

1   What is the **pitch** for a flat roof?

2   What is the **minimum height** for a roof with a pitch of **10.5/12**?

3   What is the **maximum pitch** possible for a pipe of **height 7 feet**?

4   A roof has a **vertical rise of 3 feet for every 4 feet horizontally**. What is the minimum height of a vent pipe for this roof?

5   Copy and complete this step graph.

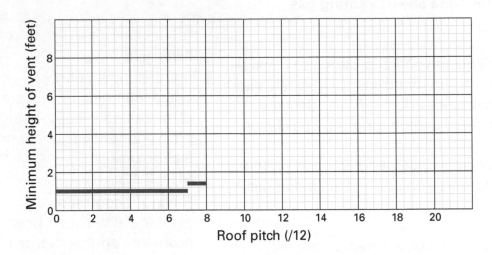

Task 2

The diagrams below are **drawn to scale**. Each square represents an area of **1 foot by 1 foot**.

Decide if the following vent pipes meet the **minimum requirements**.

If they do not, explain why.

a

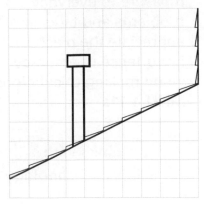

b

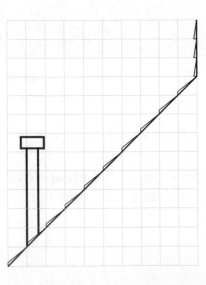

Task 3

Copy these diagrams and draw a vent of the **appropriate height** and in an **acceptable position** for each of these roofs.

a

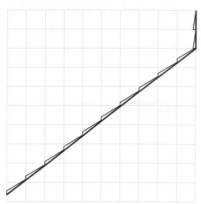

b

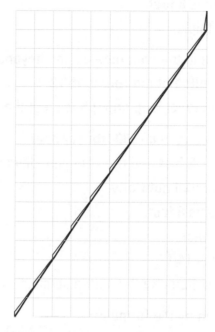

Look at the second page of the data sheet. The table shows the **maximum allowable output** of a **single gas appliance** for **different vent arrangements**.

For example, a vent pipe with a **diameter of 6 inches**, a **lateral distance of 5 feet** and a **height of 10 feet** could have a minimum of **52 000 Btu per hour** and a maximum of **280 000 for a fan-assisted system** and **188 000 for a naturally ventilated system**.

Task 4

1   An appliance has an output of **200 000 Btu per hour**.

It is fitted in a room where there is **no room for a lateral pipe** so the vent pipe rises **vertically**.

The vent pipe is **4 inches** in diameter and there will be a fan fitted.

What is the **minimum height** of the pipe?

2   A gas fitter measures for an installation. She works out that the **lateral distance possible** is **5 feet** and the **minimum height** for the vertical pipe is **15 feet**.

She intends to fit a fan to the vent pipe.

What is the **minimum diameter** of pipe, if the householder wants an appliance that produces **at least 250 000 Btu per hour**?

3 An appliance has a rating of **300 000 Btu per hour**.

The lateral distance is **5 feet** and the vent pipe diameter is **7 inches**.

The system will be vented **naturally**.

The height to the roof is **8 feet**.

The roof is **6 inches** thick.

   a By how much should the vent pipe extend **beyond the roof**?

   b Now assume the **pitch** of the roof is **10/12**.

      Will the pipe be **at least** long enough to meet the **minimum height regulations**?

4 An appliance has a rating of **80 000 Btu per hour**.

The lateral distance is **2 feet** and the vent pipe diameter is **4 inches**.

The system will be vented **naturally**.

The height to the roof is **6 feet**.

The roof is **6 inches** thick.

The pitch of the roof is **14/12**.

What is the **minimum distance** the vent pipe must extend **beyond the roof**?

5 The **flow area** is given by the formula:

$$\text{flow area} = \pi \left(\frac{D}{4}\right)^2 \text{ square inches}$$

Calculate the **flow area** for pipes of the following **diameters** (*D*). Give your answer to the **nearest whole number**.

   a 3 inches   **b** 5 inches   **c** 8 inches   **d** 9 inches

Task 5 (extension)

In the diagrams for this task, **each square** represents an area of **2 feet by 2 feet**.

1 What are the **minimum, maximum with a fan** and **maximum with natural ventilation** in **thousands of Btu per hour** for these vent systems?

a

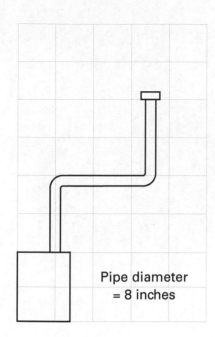

Pipe diameter
= 8 inches

b

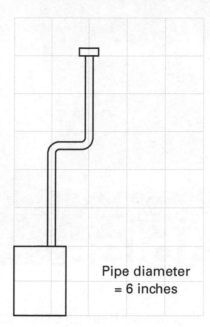

Pipe diameter
= 6 inches

2   The diagram shows an appliance with a rating of **100 000 Btu per hour**. Design a vent system for the appliance. Use a vent pipe with a diameter of **6 inches**, venting from the appliance **vertically** into the **loft space**, as shown. Complete the diagram so that the venting system **meets all the necessary regulations**. The system will be **naturally ventilated**.

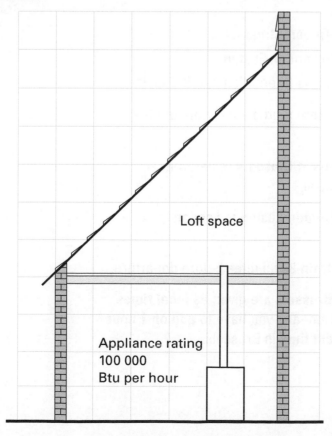

Loft space

Appliance rating
100 000
Btu per hour

Timetables

In this activity, you are going to use train timetables to plan a journey. You will need the **Data sheet: Timetables** (pp 125–126), or you could use internet searches.

CfE Outcomes

I can research, compare and contrast aspects of time and time management as they impact on me.
MNU 4-10a

Links with
Technology, Languages, Social Studies

Task

Suppose you live in **Perth** and you want to visit **Brussels**. You want to spend **three nights** there and you decide to travel by **Eurostar**.

You decide to go on a **Friday** and return on a **Monday**.

Your **outward journey** is:

- by train from Perth to Glasgow Queen Street
- by train from Glasgow Central to London Euston
- by Eurostar from London St Pancras International to Brussels.

Plan the **journey to Brussels** and the **return** journey and work out the total time spent travelling.

Allow at least **20 minutes** between trains in **Glasgow** to get from Glasgow Queen Street to Glasgow Central.

Allow at least **20 minutes** to get from **London Euston** to **London St Pancras International**.

You need to **check in for the Eurostar train 30 minutes** before departure.

The **times of arrival and departure in Brussels** are given as **local times**, and are **1 hour ahead of UK times**. This means you have to **add on 1 hour to the London time** to find the **equivalent time** in Brussels.

You could use a table like this to help you plan the trip.

	Time	Time taken
Depart Perth		
Arrive Glasgow Queen Street		
Depart Glasgow Central		
Arrive London Euston		
Depart London St Pancras		
Arrive Brussels		
	Total time taken =	

	Time	Time taken
Depart Brussels		
Arrive London St Pancras		
Depart London Euston		
Arrive Glasgow Central		
Depart Glasgow Queen Street		
Arrive Perth		
	Total time taken =	

Alcohol

Excessive consumption of alcohol is a growing problem in the United Kingdom (UK), especially among young people. It is common for young people to drink to excess, especially at weekends, in what is called binge drinking. In the UK, binge drinking is defined as consuming more than half of the weekly recommended alcohol intake in one session. It is estimated that 54% of 15–21-year-olds binge drink regularly.

This graph shows the average amount of alcohol consumed by people over 14 in the UK over the last century. The graph shows the amount of pure alcohol consumed.

Annual average alcohol consumption per person over 14 in the UK for 1900–2000

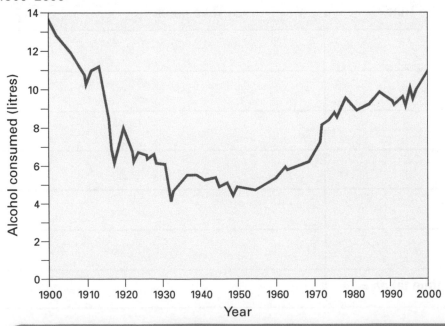

1 pint ≈ 0.57 litres   1 pound (lb) ≈ 0.45 kg
1 litre ≈ 1.76 pints   1 kilogram (kg) ≈ 2.2 lbs
1 litre = 10 dl = 100 cl = 1000 ml

Task 1

1   Suggest reasons why the consumption of alcohol was **highest** in the early years of the last century.

2   Since approximately **what year** has the consumption been increasing steadily?

3   Estimate the **average consumption** of alcohol in 2010, if the trend in the graph is continued.

4   A 333 ml bottle of German lager contains **5% alcohol by volume (ABV)**. How many bottles will it take to provide **1 litre of alcohol**?

5   A 75 cl bottle of red wine contains **14% ABV**. How many bottles will it take to provide **1 litre of alcohol**?

Task 2

In the UK, a **unit of alcohol** is defined as **10 millilitres** (or approximately **8 grams**) of **ethanol (ethyl alcohol)**.

The number of units of alcohol in a drink can be calculated by multiplying the **volume** of the drink (in **millilitres, ml**) by its **percentage ABV** and **dividing by 1000**. Thus, one pint (**568 ml**) of beer at **4% ABV** contains:

$$\frac{568 \times 4}{1000} = 2.3 \text{ units}$$

If the volume is given in **centilitres** then the formula reduces to the **volume** multiplied by the **ABV as a decimal**.

So **75 centilitres** of wine at **13% ABV** contains **75 × 0.13 = 9.75 units**

Since 1995 the UK Government has advised that:

- regular consumption of 3–4 units of alcohol a day for men and 2–3 units a day for women would not pose significant health risks

- consistently drinking four or more units a day (men) or three or more units a day (women) is not advisable.

1   How many units are there in each of the following drinks?

   a   A **75 cl** bottle of sherry with **20% ABV**

   b   A **litre** bottle of whisky with **40% ABV**

   c   A **275 ml** bottle of an alcopop with **5% ABV**

   d   A **pint** of mild beer with **2.8% ABV**

   e   A **half-pint** of strong lager with **8% ABV**

2   What is the ABV of each of these drinks?

   a   A **70 cl** bottle of cognac that contains **38 units**

   b   A **litre** bottle of vodka that contains **37.5 units** c A **75 cl** bottle of white wine that contains **8.25 units**

    d A **180 ml** bottle of barley wine that contains **1.5 units**

    e A **pint** of shandy that contains **1 unit**.

3   How many **333 ml** bottles of German lager with **5% ABV** could a **man** drink before passing the daily unit limit?

4   How many **12.5 cl** glasses of red wine with an **ABV of 14%** could a **woman** drink before passing the daily limit for women?

5   A **man** drinks **8 pints** of beer with **3.8% ABV** and **two 30 ml** shots of whisky with **40% ABV** on a Friday evening.

    a How many **units** of alcohol does he consume?

    b How many **centilitres** of alcohol does he consume?

Task 3

In the UK, and in most other countries, drink–driving is considered to be a serious offence. The limit for driving in the UK is **80 ml of alcohol per litre of blood** or **8% blood alcohol**. Alcohol inhibits reaction times and causes drivers to feel over-confident, which leads to many accidents.

Look at **Data sheet: Alcohol** (p 127). You will need to refer to it to answer these questions.

1   A **woman** weighing **140 lbs** drinks a **pint** of beer with **4% ABV**. Would she be **legally intoxicated**?

2   A **woman** drives to a restaurant in her car. She weighs **160 lbs**. During the meal, which takes **2 hours** she drinks two **125 ml** glasses of white wine with **11% ABV**. Is she safe to drive home?

3   Over the course of a Friday evening from **8 pm to 11 pm** a **man** weighing **160 lbs** drinks **4 pints** of beer with **3.8% ABV**. Work out his **probable blood alcohol percentage**.

4   Four teenagers – two girls and two boys – obtain a **2-litre** bottle of cheap cider with **7.5% ABV**. The girls weigh **100 lbs each** and the boys weigh **120 lbs each**. They share the cider equally. Work out the **probable blood alcohol percentage** of each of them.

5   Every Sunday lunchtime a man, weighing **160 lbs**, drives his family to a local pub for lunch. He drinks **1 pint** of beer with **3.5% ABV** and a **250 ml** glass of wine with **12% ABV**. The meal lasts **90 minutes**. After the meal the family go for a walk. How long should the walk last if he is to be **safe enough to drive home**?

Task 4 (extension)

Design a poster warning of the effects of drink–driving.

Garden designer

You have probably seen TV shows where people go to a house where the garden looks awful and, with some effort, change it into a something completely different.

Imagine that you work as a garden designer.

You have been sent to a new house to give an estimate for a job.

Here is a sketch of the garden area.

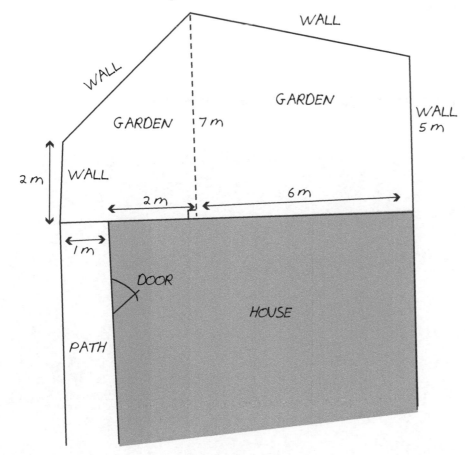

Task 1

Using **centimetre-squared paper** and a **scale of 1 cm to represent 1 metre**, make two accurate **scale drawings** of the garden shown in the sketch.

You will use these in **Task 2** and **Task 3**.

Task 2

The home-owner wants the garden design to include the **following features**.

There should be:

- a **path** along the **wall** of the house – you can choose how wide it is
- a **seating area** close to the house
- a **lawn** that has an **area of at least 16 m²**
- a **circular pond** of **radius 1 metre**
- a **flower bed** that is **half a metre wide** and **between 3 metres and 4 metres long**
- **two trees** that have to be **more than 2 metres from the house wall** and **at least 3 metres apart**
- a **rockery** next to one wall – its **area** has to be **between 2 m² and 5 m²**.

Design the garden.

Include all the **measurements** on your diagram.

Task 3

Using exactly the **same rules as in Task 2**, redesign the garden to give the homeowner a **choice of designs**.

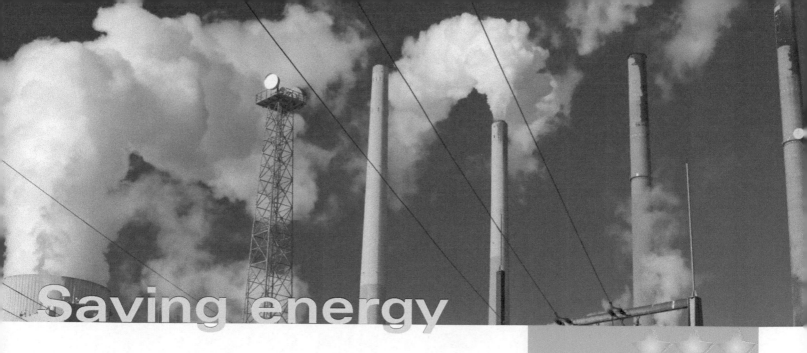

Saving energy

The effects of global warming and increasing costs of energy are likely to affect everyone. In this activity, a family is trying to work out how they can save money and reduce carbon emissions.

Your role is that of a consultant who can help the family take sensible decisions. This is your business card.

Mr and Mrs Patel have asked you to give them advice about saving money, especially on their annual gas and electricity bills.

I. M. Green

Energy-saving consultant

*Call me to save money
on your energy bills*

01234–555–6789
imgreen@alo.co.uk

This is a plan of their bungalow. Each square represents 1 square metre.

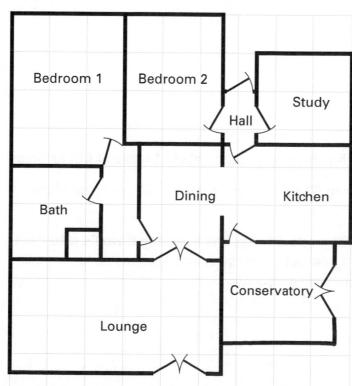

Bedroom 1

Bedroom 2

Study

Hall

Bath

Dining

Kitchen

Conservatory

Lounge

Room area (m^2)	Minimum Btu (per hour)
$6 \leqslant$ area < 8	1800
$8 \leqslant$ area < 10	2100
$10 \leqslant$ area < 12	2400
$12 \leqslant$ area < 14	2700
$14 \leqslant$ area < 18	3000

This table shows the minimum radiator output required per hour, in British thermal units (Btu), for rooms of various areas.

Task 1

The Patels have a radiator in each of their rooms **except the hall**.

Calculate an approximate value for the **minimum Btu (per hour)** they will need for each room.

They run their heating from 6 am to 8 am and from 4 pm to 10.30 pm on weekdays and from 6 am to 10.30 pm on Saturdays and Sundays. Calculate their **approximate annual Btu use**.

> One and a half million Btu $\approx$ 60 kg of carbon dioxide gas (CO_2)

How many tonnes of CO_2 does the Patels' heating system produce each year?

Task 2

The Patels' **annual bills** are:

- **electricity £540**
- **gas £720**
- **water £640**.

Assume that their use of energy other than heating produces about **1 tonne of CO_2 per year**.

The table opposite shows the **cost**, **saving (£) per year** and **saving in CO_2 per year** of the various energy-saving methods.

Write a report for the Patels, outlining the **costs**, **savings** and **how long it would take** them to get their money back if they bought a new set of energy-saving appliances, for example. You **do not** have to **include** every energy-saving measure.

Action	Annual CO$_2$ saving	Annual cost saving
Turn down the heating by 1°C	330 kg	£40
Switch off electrical goods when not in use	153 kg	£40
Turn off the tap		£7 per tap
Dry clothes on the line	311 kg	£15
Draught proofing	150 kg	£25 (cost of DIY: £90)
Use shower instead of bath		£22 per person
Set hot water to 60°C	145 kg	£10
Use energy-saving bulbs	40 kg	£9 (cost: 49p each)
Use Savaplug on fridge	100 kg	Save £12 (cost: £25)
Fill kettle with only as much water as required	48 kg	
Replace old appliances with energy-efficient ones	85 kg	£70 (cost approx: £1000)
Double-glaze windows	720 kg	£110 (cost approx: £400 per room)
Fill up the dishwasher before using		Save £15
Insulate loft	1 tonne	£150 (cost of DIY: £250)
Install solar electric power	1.2 tonne	Half of the electricity bill (cost: £6000 to £15000)
Fit foil behind radiators	51 kg	(cost: £15)
Install wind power generator	250 kg	One-third of the electricity bill (cost: from £1500)
Insulate cavity walls	800 kg	£120 (cost: £500)

Ways to save energy and the planet – annual savings for each energy-savings measure

Task 3 (extension)

The Patels currently drive an old car that produces **160 g of CO$_2$ per kilometre**. They travel **16 000 km per year** by car. The car does about **7.5 kilometres to a litre** of fuel. Use the internet to research the **least polluting** cars. Advise the Patels about whether it will be worth their while buying a newer model.

Rugby numbers

In a game of Rugby Union you score points as follows.

Try	5 points
Conversion	2 points
Drop goal	3 points
Penalty goal	3 points

You can only attempt a conversion after you have scored a try.
A converted try is worth 7 points (5 + 2).
An unconverted try is worth 5 points.

Here is the result of a rugby match played on 29 January 2011.

Melrose	45 points	7 tries 5 conversions 0 penalties
Hawick	23 points	3 tries 1 conversions 2 penalties

Task 1

You may choose to work **in pairs** for this task.

Check that the **score** shown above is **correct**.

Can you find a **different way** in which **Hawick** could have scored **23 points**?

Task 2

You may choose to work **in pairs** for this task.

Explain why it **impossible** for a team to score **exactly 4 points**.
What other **total numbers of points** are **impossible**?

Some totals can be reached in **more than one way**.
Explore the **different ways** in which **total scores** can be achieved.

> The way points are scored has changed over the years.
>
> **Until 1971** a **try** was worth **3 points**.
>
> From **1971 to 1992** a **try** was worth **4 points**.
>
> A **try** has been worth **5 points** since **1992**.

Task 3

You may choose to work **in pairs** for this task.

What would the **scores** in the **Melrose vs Hawick match** have been if a
try had been **worth 4 points** or **3 points**?
Would the **winner** have been **different** in **either case**?

Task 4

You may choose to work **in pairs** for this task.

Can you think of some results for which the **winner**
using a **5-point-try** scoring system would **lose** using a
3-point-try scoring system?

Can you **generalise** your findings?

Why do you think that the **number of points** for a **try**
was **increased**?

Follow that car

Cars have number plates so that they can be identified uniquely.

The registration letters on a car show where it was first registered.

Do cars tend to stay where they were registered or do they move around the country?

In this activity you will investigate what information you can find out about a car just from its number plate.

You will need the **Data sheet: Follow that car** (pp 128–129).

Task 1

Think of a car that you know that was registered **after 1 September 2001**.
Find out all you can from its **number plate**.
You could use the internet to help you.

Task 2

Think up one or two questions you could ask someone about **car registration codes** to see if that person understands how to **decode** them.

Task 3

Investigate cars **in your area** that have **new-style number plates**.

What **hypotheses** could you test?

What **questions** could you ask?

Choose a topic about car registration to **investigate**.

Write a **brief plan**, including the **data** you need to collect and what you will do with it.

Task 4

Carry out your **enquiry** from Task 3.

Present your results in any way that you think is suitable.

Bike race

Bicycle track racing is a sport in which Britain does well. It requires very high levels of fitness and great dedication in training.

You will need the **Data sheet: Bike race** (pp 131–133) for this activity.

Task 1

Study the result of the **one-kilometre time trial** in the **World Cycling Championships**.

Who **won**? What was his **winning time**?

Write down five interesting facts from the data.

Task 2

Look at the details of the **medal winners**.

These are the riders who came **first, second** and **third**.

Analyse their performances and, in particular, compare their **times** and **positions** as the **race progressed**.

Draw up appropriate **graphs** and **charts** to do this.

Think carefully about the **best type of graph or chart** to use.

Include a **brief description** of any **interesting features**.

Task 3

From the data you have been given, you could work out the time for a rider to travel **each 125 metres** of the race. How do you think these will vary?

Do some **calculations** to see if you are correct.

It may be **different** for different riders.

You could show your times on a **chart or graph**.

Task 4

Is it possible to work out the **exact speed** of a rider at any time?

Use **average speeds** to **estimate** the **fastest speeds** of some riders.

How much difference is there between the **fastest** and the **slowest** riders?

> **Useful information**
>
> $$\text{Average speed} = \frac{\text{distance travelled}}{\text{time taken}}$$
>
> 1 km/h = 0.62 mph
>
> 1 m/s = 2.24 mph
>
> In 2005 the **world record** distance **cycled in one hour** from a **standing start** was set by **Ondrej Sosenka**. He travelled **49.700 km** (30.88 miles).

Tile that wall

Wall tiles are usually placed according to a pattern that looks attractive.

One of the ways of creating a pleasing pattern is to use symmetry.

CfE Outcomes

I can illustrate the lines of symmetry for a range of 2D shapes and apply my understanding to create and complete symmetrical pictures and patterns.
MTH 2-19a/MTH 3-19a

Links with
Expressive Arts

Task 1

Suppose you have four tiles like this:

Can you arrange them to make a **symmetrical image**?

Can you do it in **different ways**?

Think about **reflection symmetry** or **rotation symmetry** or **translations**.

Task 2

Work in pairs for this task.

Imagine you are a designer for a company that makes tiles.

Your brief is to **design a square tile** that can make an interesting **symmetrical pattern** to cover a wall.

There are some restrictions which will help keep down the cost.

- Your design should be **simple to draw**.
- You can only use **white** and **one other colour**.

Design a suitable tile.

Show how your tile can be used to make an **interesting symmetrical pattern**.

Be prepared to **describe the symmetries** in your pattern.

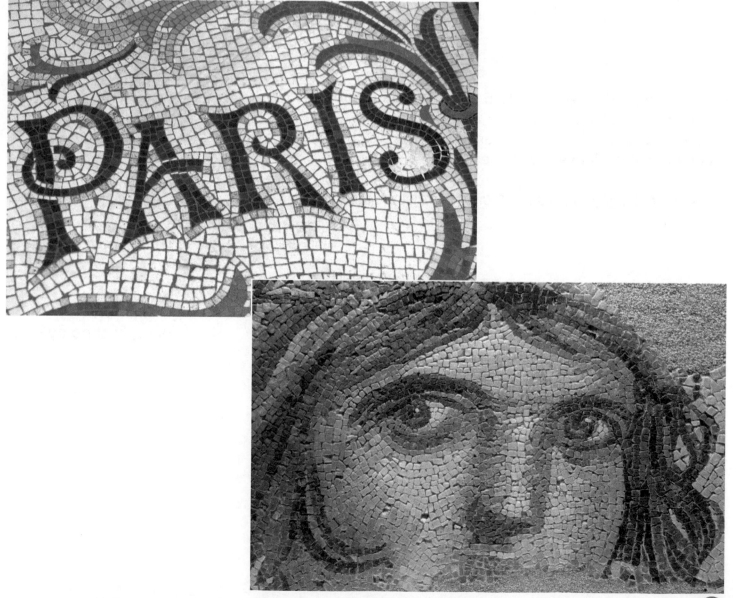

Body mass index

The **body mass index** (**BMI**) is a measure used to show if an adult is at a healthy weight.

Here is a formula for calculating BMI.

$$\text{BMI} = (\text{weight in kg}) \div (\text{height in m})^2$$

A person with BMI between 18.5 and 25 is at a healthy weight.
A person with BMI less that 18.5 is underweight.
A person with BMI between 25 and 30 is overweight.
A person with BMI over 30 is obese.

This does not apply to children. A different method is used for them.

Task 1

A woman is **1.63 m tall** and **weighs 43.5 kg**.
She is worried that she is **overweight**. What advice can you give her?

Task 2

Work in **pairs**.

A man with a **BMI of 25** is on the **borderline** of being **overweight**. Work out some **possible heights** and **weights** for him.

Task 3

Look at these four people.

Here are the **heights** and **weights** of four people, A, B, C and D, but not in any particular order.

Height (m)	1.75	1.20	1.65	1.60
Weight (kg)	53	31	57	80

Decide whether each of them is **underweight**, **healthy**, **overweight** or **obese**.

Task 4

Work in groups of **four**.

Produce a **poster** that **adults** can use to decide whether they are at a **healthy weight**.

> Some people only know their **height** in **feet** and **inches** and their **weight** in **stones** and **pounds**.
>
> 1 foot = 30.5 cm or 0.305 m   12 inches = 1 foot   1 inch = 2.54 cm
>
> 1 stone = 6.35 kg   14 pounds = 1 stone   1 pound = 0.454 kg

Task 5 (extension)

My dad is **6 feet tall** and **weighs 15 stone**.

He says his **BMI** = $15 \div 6^2 = 15 \div 36 = 0.417$.

That can't be right! What has gone wrong? What is the correct answer?

Task 6 (extension)

Scott Lawson is a rugby player. He was in the Scotland squad selected for the Rugby World Cup in 2011.

When he played for Scotland he was **1.73 m tall** and **weighed 102 kg**.

Do you think he was overweight?

Green travel

Carbon dioxide is one of the greenhouse gases that contribute to global warming.

The Government has put cars into thirteen bands, based on their carbon dioxide emissions.

Band	CO$_2$ emission (g/km)
A	Up to 100
B	101–110
C	111–120
D	121–130
E	131–140
F	141–150
G	151–165
H	166–175
I	176–185
J	186–200
K	201–225
L	226–255
M	Over 255

For example, a Range Rover gives out about 330 g/km and a Ford Fiesta gives about 130 g/km, depending on the model. Cars with higher emissions pay more road tax. Cars in band A pay none.

You can find ready reckoners to give information about cars' carbon dioxide emissions and the tax bands on the internet.

Figures for other types of transport can also be found on the internet or in the table opposite.

Task 1

Explain what **g/km** means.

What do you think the **CO₂ emission rate** is for any cars in your family?
If your family does not have a car, think about a car belonging to someone you know.

Task 2

Suppose you came to school **by car** every day.

How much **carbon dioxide** would the car **emit** into the atmosphere **in a year**, carrying you **to and from school**?

Task 3

Suppose you work as a **green travel consultant**.

A client who **travels regularly** to various parts of **Great Britain** has asked for your advice about the **CO₂ emissions** for the different ways she travels.

She could travel by **car, train, coach** or **plane**.

Your task is to give her some **comparative figures** for **CO₂ emissions**, for travelling from **London to Edinburgh**.

Then find similar figures for **typical journeys** in your **local area**.

The distance from **London to Edinburgh**, by road, is **661 km**.

Kilograms of CO_2 per passenger per kilometre (public transport)	
Air journeys	0.1753
Bus journeys	0.1073
Coach journeys	0.029
Light Rail journeys	0.078
Tyne & Wear Metro journeys	0.1207
Docklands Light Rail journeys	0.074
Croydon Tramlink journeys	0.042
Manchester Metrolink journeys	0.0421
Rail journeys	0.0602
London Underground journeys	0.065
Ferry journeys	0.1152

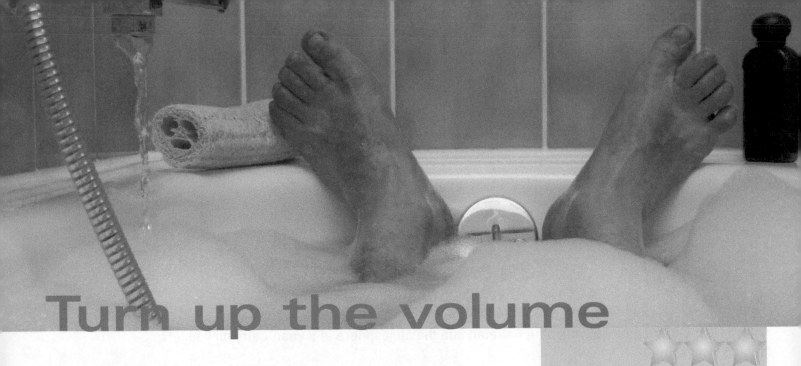

Turn up the volume

Volume is a measure of how big an object is. It is easy to measure the volume of boxes but in this activity you will find out how to measure the volume of irregular shapes such as people.

Task 1

How can you find the **volume** of a **cereal box**?

What **units** will you use for the answer?

How **accurately** can you give the answer?

Task 2

Do this activity in pairs.

Can you find the **volume** of a **person**?

What **methods** can you think of for doing this?

Your task is to choose a method and estimate the volume of a person as accurately as you can.

You should be able to explain your method to someone else.

Task 3 (extension)

In 1946 Robert Earl Hughes was the **world's fattest man** and he still holds the world record for the largest **chest measurement: 10 feet 4 inches**.

Source: *Guinness Book of Records*

When he died he was buried in a **coffin** the size of a **piano case**.

Can you estimate the volume of Robert Earl Hughes and compare it to the volume of a 'normal' pupil?

Useful information:

1 foot is approximately 30 cm

1 inch is approximately 2.5 cm

Task 4 (extension)

Can you estimate the **volume** of a **baby**?

How did you do it?

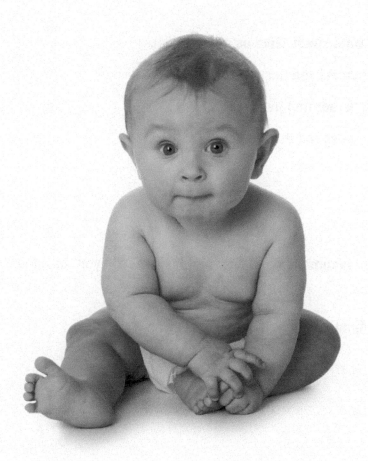

Give us a job

Getting a job can seem like the answer to all your problems. You have money to spend and things to spend it on. The reality is often tougher than people expect.

Task 1

Choose a job from **Data sheet: Give us a job** (p 134).

Imagine you have **just started** this job.

How much will you be **earning per week**?

How would you like to **spend this money**?

Task 2

You will have to pay **income tax** and **National Insurance** in your new job.

How much will you have to pay?

Will you have to **adjust** your **spending plans**?

Task 3

You have decided to live **away from home**.

You want to **share a house** or a **flat** with some friends.

Do some **research** to find out how much this will **cost** you.

How much money have you **got left** now?

How will you **spend it**?

Task 4

Have you included **food** in your **spending plans**?

Estimate how much you will spend **in a week** on **food**.

How much have you got **left** now?

Task 5

What **other things** will you have to pay for?

Try to make an **estimate** of those costs.

Have you got **any money left**?

Task 6 (extension)

Many people like to have a **car**.

Your granddad has offered to **give you his car** because he cannot drive any more.

Can you **afford** to **run it** on your **salary**?

Populations

Is the size of a country is directly related to the number of people who live there? In this activity, you will investigate.

There are many websites where you can find out about the populations and the areas of the countries of the world.

> The **population density** of a country is calculated as
>
> $$\text{population density} = \frac{\text{population}}{\text{area}}$$
>
> For example, a 2010 estimate for the population of Scotland was 5.22 million. The area is 78 772 square kilometres.

Task 1

1   Check that the population density of Scotland in 2010 is about 66 people per km².

2   The population density of England is about **400 people per km²**.
    Based on the two population densities, which of these statements must be true?
    A   England has a bigger population than Scotland
    B   England has a bigger area than Scotland
    C   English cities are more crowded than Scottish cities
    D   England is more crowded than Scotland
    E   England is a nicer place to live than Scotland

Task 2

Work in pairs for this task if you wish.

Refer to **Data sheet: Populations** (pp 135–138), which shows the **populations** and the **areas** of all the countries of the world. Think of some **questions** you could ask. Here are some examples.

- Do the **largest countries** have the **biggest populations**?

- Which countries in **Europe** have **similar population densities** to that of the **UK**?

- How much do the populations of **African** countries vary?

- Think of some **more questions** you could answer, using the **Data sheet**.

- How could you **change** each question into a **hypothesis to test**?

Task 3 (investigation)

Choose a **question or hypothesis** from Task 2.

Use the **Data sheet** to find the information you will need to **answer your question** or **test your hypothesis**.

Carry out the **investigation** and write down your **conclusion**. Include **evidence** such as **charts** or **graphs** to support your conclusion.

Be prepared to **present your findings** in a **discussion** with the rest of the class.

Data sheet: Scotland

Population:

- approximately 5.22 million.

- 17% of population aged under 16.

- 87% of people living in Scotland in 2001 were born in Scotland.

- 8.2% of people living in Scotland in 2001 were born in England.

- 4.8% of people living in Scotland in 2001 were born in other UK countries or countries outside the UK.

Facts

- Scotland has an area of 78 772 square kilometres (about 30 414 square miles).

- 7.2% of Scotland is designated as National Park.

- The capital of Scotland is Edinburgh.

- There are 10 000 km of coastline (3,900 km on the mainland and 6,100 km on the islands).

- Ben Nevis, at 1344 m, is the highest mountain in Scotland and higher than any other peak in the UK.

- One in 3 Scottish MSPs are women.

- The coldest day on record in Scotland was 30 December 1995 when the temperature was –27.2 °C at Altnaharra in the Highlands.

- The warmest day on record in Scotland was 9 August 2003 when the temperature was 32.9 °C at Greycrook in the Borders.

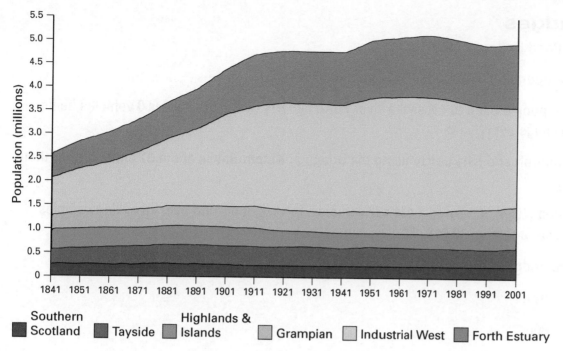

Southern Scotland | Tayside | Highlands & Islands | Grampian | Industrial West | Forth Estuary

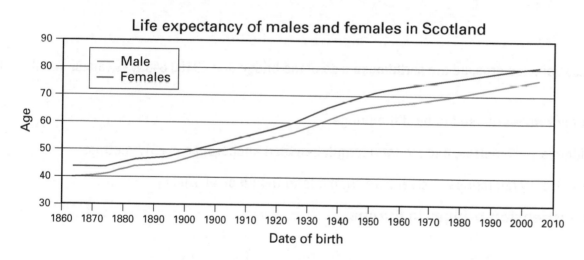

Data sheet: Bridges

Work started on the Forth Road Bridge in **1958**.

The Forth Road Bridge was opened by the Queen in **September 1964**.

Before the bridge was built, people used to cross the river Forth by ferry, with about **800 000** vehicles being carried on **40 000** ferry crossings every year.

The **distance** between Edinburgh and Kirkcaldy using the bridge at **Kincardine** is about **31 miles** further than using the Forth Road Bridge.

In the year the bridge opened (1964), it carried **1 434 422** vehicles. In 1978, it carried over 10 million vehicles per year for the first time, with a total of **10 018 488** crossings.

In 1997, **toll charges** for the bridge were:

Motor cycles	free
Cars	80p
Buses	£1.40
Lorries	£2.00

In 2000, a total of **11 084 253** vehicles crossed the **northbound** side of the bridge and £9 150 848 was raised in toll fees.

125 000 cubic metres of **concrete** were used to build the bridge.

The **main span** of the bridge is 1006 metres, and the **total length** between anchorages is 2517 metres.

The **main cables** are made from 7720 tonnes of steel wire, with a **total length** of 49 570 km.

Longest single-span suspension bridges in Europe.

Name	Country	Span (metres)	Year of opening
Great Belt Bridge	Denmark	1624	1998
Humber Bridge	England	1410	1981
Höga Kusten	Sweden	1210	1997
Ponte 25 de Abril	Portugal	1013	1966
Forth Road Bridge	Scotland	1006	1964
Severn Bridge	England/Wales	988	1966
Askøy Bridge	Norway	850	1992
Stord Bridge	Norway	677	2001
Tancarville Bridge	France	608	1959
New Little Belt Bridge	Denmark	600	1970

Data sheet: Planning a bedroom

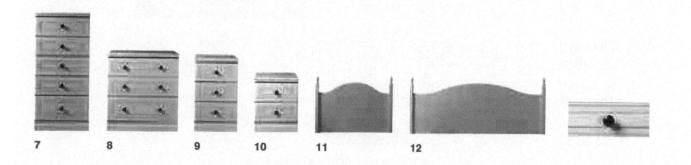

1.	**3 door mirror wardrobe***	H1873 x W1141x D525mm	5017860716557	**£400**
2.	**2 door wardrobe**	H1873 x W760 x D525mm	5017860716496	**£250**
3.	**2 door combi wardrobe**	H1873 x W760 x D525mm	5017860716335	**£300**
4.	**Dressing table**	H759 x W972 x D406mm	5017860716458	**£275**
5.	**5 drawer wide chest**	H1027 x W766 x D406mm	5017860716373	**£225**
6.	**7 drawer wide chest**	H822 x W1147 x D406mm	5017860716618	**£275**
7.	**5 drawer tall chest**	H822 x W1147 x D406mm	5017860716397	**£225**
8.	**3 drawer wide chest**	H822 x W766 x D406mm	5017860716359	**£200**
9.	**3 drawer bedside chest**	H759 x W450 x D406mm	5017860716472	**£125**
10.	**2 drawer bedside chest**	H554 x W448 x D406mm	5017860716311	**£100**
11.	**Single headboard**	H587 x W914 x D70mm	5017860716410	**£100**
12.	**Double Headboard**	H587 x W1524 x D70mm	5017860716434	**£150**

Everything here is ready-assembled and delivered direct to your home.*

* With the exception of the 3 door wardrobes which are flat-packed for self-assembly to allow access into your home.

Bed sizes

small single	75 cm x 190 cm	double	135 cm x 190 cm
single	90 cm x 190 cm	king	150 cm x 200 cm
super single	105 cm x 190 cm	super king	180 cm x 200 cm
three-quarter (small double)	120 cm x 190 cm		

Data sheet: Stopping distances

This diagram shows the **thinking distance** and **braking distance** for a driver to react and stop in an emergency.

The **thinking distance** is the distance the car travels **while the driver's brain reacts**; this is before the driver starts to apply the brake to the car.

The **braking distance** is the distance the car will travel before it comes to a stop.

The **braking distance** is approximate and represents a **maximum**. In reality, the braking distance may be less.

One **car length** is **4 metres** so at **20 mph** the total **braking distance** is **12 metres** or **3 car lengths**.

It would therefore be sensible for a driver to leave **3 car lengths** between their car and the car in front to be sure that there is sufficient distance to think and brake in an emergency.

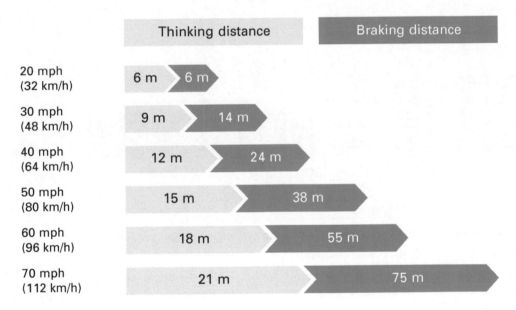

Emergency braking from 60 mph

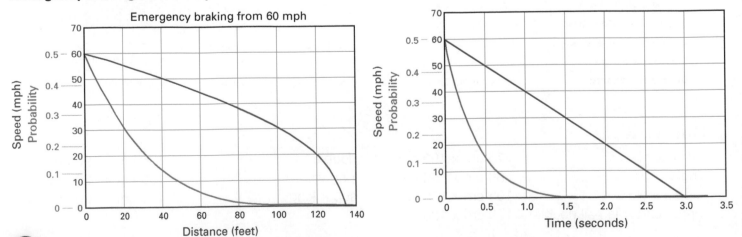

Data sheet: Venting gas appliances

Roof pitch	Minimum height of vent (feet)
Flat to 7/12	1.0
Over 7/12 up to 8/12	1.5
Over 8/12 up to 9/12	2.0
Over 9/12 up to 10/12	2.5
Over 10/12 up to 11/12	3.25
Over 11/12 up to 12/12	4.0
Over 12/12 up to 14/12	5.0
Over 14/12 up to 16/12	6.0
Over 6/12 up to 18/12	7.0
Over 18/12 up to 20/12	7.5
Over 20/12 up to 21/12	8.0

Roof pitch and minimum height of vent

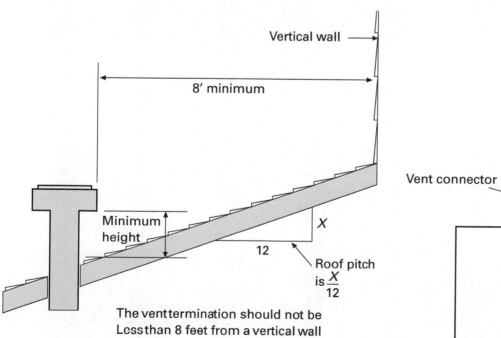

The vent termination should not be Less than 8 feet from a vertical wall

Diagram showing the position of the vent termination relative to the vertical wall

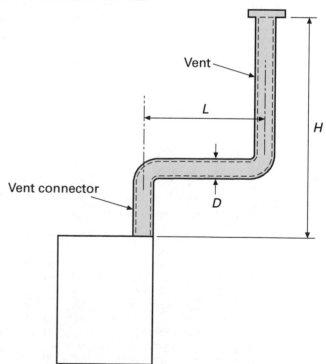

Diagram for a single gas appliance showing the diameter, *D*, lateral distance, *L*, and height, *H*, for vent systems

Vent table: capacity of Type B double-wall vents with Type B double-wall connectors serving a single category 1 appliance
FAN = fan assisted extraction. NAT = natural extraction

VENT TABLES

Capacity of Type B Double-Wall Vents with Type B Double-Wall Connectors
Serving a Single Category I Appliance

TABLE 1

Height H (ft)	Lateral L (ft)	3″ FAN Min	3″ FAN Max	3″ NAT Max	4″ FAN Min	4″ FAN Max	4″ NAT Max	5″ FAN Min	5″ FAN Max	5″ NAT Max	6″ FAN Min	6″ FAN Max	6″ NAT Max	7″ FAN Min	7″ FAN Max	7″ NAT Max	8″ FAN Min	8″ FAN Max	8″ NAT Max	9″ FAN Min	9″ FAN Max	9″ NAT Max
6	0	0	78	46	0	152	86	0	251	141	0	375	205	0	524	285	0	698	370	0	897	470
	2	13	51	36	18	97	67	27	157	105	32	232	157	44	321	217	53	425	285	63	543	370
	4	21	49	34	30	94	64	39	153	103	50	227	153	66	316	211	79	419	279	93	536	362
	6	25	46	32	36	91	61	47	149	100	59	223	149	78	310	205	93	413	273	110	530	354
8	0	0	84	50	0	165	94	0	276	155	0	415	235	0	583	320	0	780	415	0	1006	537
	2	12	57	40	16	109	75	25	178	120	28	263	180	42	365	247	50	483	322	60	619	418
	5	23	53	38	32	103	71	42	171	115	53	255	173	70	356	237	83	473	313	99	607	407
	8	28	49	35	39	98	66	51	164	109	64	247	165	84	347	227	99	463	303	117	596	396
10	0	0	88	53	0	175	100	0	295	166	0	447	255	0	631	345	0	847	450	0	1096	585
	2	12	61	42	17	118	81	23	194	129	26	289	195	40	402	273	48	533	355	57	684	457
	5	23	57	40	32	113	77	41	187	124	52	280	188	68	392	263	81	522	346	95	671	446
	10	30	51	36	41	104	70	54	176	115	67	267	175	88	376	245	104	504	330	122	651	427
15	0	0	94	58	0	191	112	0	327	187	0	502	285	0	716	390	0	970	525	0	1263	682
	2	11	69	48	15	136	93	20	226	150	22	339	225	38	475	316	45	633	414	53	815	544
	5	22	65	45	30	130	87	39	219	142	49	330	217	64	463	300	76	620	403	90	800	529
	10	29	59	41	40	121	82	51	206	135	64	315	208	84	445	288	99	600	386	116	777	507
	15	35	53	37	48	112	76	61	195	128	76	301	198	98	429	275	115	580	373	134	755	491
20	0	0	97	61	0	202	119	0	349	202	0	540	307	0	776	430	0	1057	575	0	1384	752
	2	10	75	51	14	149	100	18	250	166	20	377	249	33	531	346	41	711	470	50	917	612
	5	21	71	48	29	143	96	38	242	160	47	367	241	62	519	337	73	697	460	86	902	599
	10	28	64	44	38	133	89	50	229	150	62	351	228	81	499	321	95	675	443	112	877	576
	15	34	58	40	46	124	84	59	217	142	73	337	217	94	481	308	111	654	427	129	853	557
	20	48	52	35	55	116	78	69	206	134	84	322	206	107	464	295	125	634	410	145	830	537
30	0	0	100	64	0	213	128	0	374	220	0	587	336	0	853	475	0	1173	650	0	1548	855
	2	9	81	56	13	166	112	14	283	185	18	432	280	27	613	394	33	826	535	42	1072	700
	5	21	77	54	28	160	108	36	275	176	45	421	273	58	600	385	69	811	524	82	1055	688
	10	27	70	50	37	150	102	48	262	171	59	405	261	77	580	371	91	788	507	107	1028	668
	15	33	64	NR	44	141	96	57	249	163	70	389	249	90	560	357	105	765	490	124	1002	648
	20	56	58	NR	53	132	90	66	237	154	80	374	237	102	542	343	119	743	473	139	977	628
	30	NR	NR	NR	73	113	NR	88	214	NR	104	346	219	131	507	321	149	702	444	171	929	594
50	0	0	101	67	0	216	134	0	397	232	0	633	363	0	932	518	0	1297	708	0	1730	952
	2	8	86	61	11	183	122	14	320	206	15	497	314	22	715	445	26	975	615	33	1276	813
	5	20	82	NR	27	177	119	35	312	200	43	487	308	55	702	438	65	960	605	77	1259	798
	10	26	76	NR	35	168	114	45	299	190	56	471	298	73	681	426	86	935	589	101	1230	773
	15	59	70	NR	42	158	NR	54	287	180	66	455	288	85	662	413	100	911	572	117	1203	747
	20	NR	NR	NR	50	149	NR	63	275	169	76	440	278	97	642	401	113	888	556	131	1176	722
	30	NR	NR	NR	69	131	NR	84	250	NR	99	410	259	123	605	376	141	844	522	161	1125	670
100	0	NR	NR	NR	0	218	NR	0	407	NR	0	665	400	0	997	560	0	1411	770	0	1908	1040
	2	NR	NR	NR	10	194	NR	12	354	NR	13	566	375	18	831	510	21	1155	700	25	1536	935
	5	NR	NR	NR	26	189	NR	33	347	NR	40	557	369	52	820	504	60	1141	692	71	1519	926
	10	NR	NR	NR	33	182	NR	43	335	NR	53	542	361	68	801	493	80	1118	679	94	1492	910
	15	NR	NR	NR	40	174	NR	50	321	NR	62	528	353	80	782	482	93	1095	666	109	1465	895
	20	NR	NR	NR	47	166	NR	59	311	NR	71	513	344	90	763	471	105	1073	653	122	1438	880
	30	NR	NR	NR	NR	NR	NR	78	290	NR	92	483	NR	115	726	449	131	1029	627	149	1387	849
	50	NR	NR	NR	NR	NR	NR	NR	NR	NR	147	428	NR	180	651	405	197	944	575	217	1288	787

Data sheet: Timetables

Perth to Glasgow
Weekdays and Saturdays

Perth	0812	0915	1014	1137	1314	1514	1655	1914	2118	2230
Gleneagles	0826			1152			1710			2246
Stirling	0843	0943	1043	1212	1343	1543	1728	1943	2146	2304
Glasgow Queen St.	0915	1014	1114	1245	1414	1614	1809	2015	2220	2339

Sundays

Perth	1158	1450	1705	1830	2105
Gleneagles	1214	1504	1719	1846	2119
Stirling	1234	1522	1737	1904	2138
Glasgow Queen St.	1405	1556	1812	1937	2215

Glasgow to Perth
Weekdays and Saturdays

Glasgow Queen St.	0706	0841	1011	1141	1341	1611	1641	1741	1941
Stirling	0733	0907	1036	1207	1407	1637	1707	1807	2007
Gleneagles			1054			1700		1827	
Perth	0804	0936	1113	1236	1436	1718	1736	1843	2036

Sundays

Glasgow Queen St.	0938	1345	1440	1810
Stirling	1012	1412	1509	1839
Gleneagles	1030	1429	1528	1858
Perth	1047	1446	1545	1915

London to Glasgow
Mondays to Fridays

London Euston	0730	0830	0930	1030	1130	1230	1330	1430	1530	1630	1730	1830	1930
Lancaster	0954	1052	1153	1300	1354	1502		1700	1754		1954	2054	2158
Carlisle	1046	1146	1246	1357	1446		1646		1847		2046	2149	2250
Glasgow Central	1201	1301	1401	1524	1600		1801			2038	2201	2304	0006

Saturdays and Sundays

London Euston	0730	0830	0930	1030	1130	1230	1330	1430	1530	1630	1730	1830
Lancaster	0955	1054	1154	1254	1353	1454	1554	1654	1752	1854	1954	2054
Carlisle	1048	1146	1247	1346	1447	1545	1646	1745	1842	1945	2046	2145
Glasgow Central	1202	1301	1402	1501	1601	1701	1801	1901	2000	2101	2201	2301

Glasgow to London
Mondays to Fridays

Glasgow Central	0540	0630	0737	0840	0923	1140	1240	1440	1640	1840
Carlisle	0649	0746	0849	0949	1045	1249	1349	1549	1751	1949
Lancaster	0738	0838	0938	1038	1138	1326	1438	1638	1844	2038
London Euston	1012	1112	1212	1312	1412	1603	1712	1913	2124	2338

Saturdays and Sundays

Glasgow Central	0540	0735	0840	0940	1040	1140	1240	1340	1440	1540	1640	1740
Carlisle	0649	0849	0949	1049	1149	1249	1349	1449	1549	1649	1752	1852
Lancaster	0738	0938	1038	1138	1238	1338	1438	1538	1638	1738	1841	1944
London Euston	1012	1212	1312	1412	1512	1612	1712	1812	1912	2015	2138	2244

Eurostar
London to Brussels

LONDON St Pancras Int	EBBSFLEET International	ASHFORD International	CALAIS Fréthun	LILLE Europe	BRUSSELS Midi/Zuid	Train no.
Monday to Friday						
07:34	-	-	-	09:54	10:33	9112
08:27	08:45	-	-	10:54	11:33	9120
10:57	11:15	-	12:59	13:31	14:12	9126
12:57	13:15	-	-	15:24	16:05	9132
14:34	-	-	-	16:54	17:33	9138
17:04	-	-	-	19:24	20:05	9148
18:34	-	-	20:29	20:59	21:37	9154
19:34	-	-	-	21:54	22:33	9158

LONDON St Pancras Int	EBBSFLEET International	ASHFORD International	CALAIS Fréthun	LILLE Europe	BRUSSELS Midi/Zuid	Train no.
Saturday and Sunday						
07:57	-	08:28	-	10:30	11:08	9114
08:57	09:15	-	10:59	11:30	12:08	9116
11:57	12:15	-	-	14:24	15:05	9130
14:34	-	-	-	16:54	17:33	9138
17:04	-	-	-	19:24	20:05	9148
18:25	-	18:55	-	20:51	21:30	9154
19:34	-	-	21:29	22:01	22:39	9158

Brussels to London

BRUSSELS Midi/Zuid	LILLE Europe	CALAIS Fréthun	ASHFORD International	EBBSFLEET International	LONDON St Pancras Int	Train no.
Monday to Friday						
06:51	07:28	08:02	-	-	07:56	9109
08:05	-	-	-	-	08:59	9113
09:59	10:35	-	-	-	10:56	9121
12:29	13:05	-	-	13:15	13:33	9129
14:29	15:05	-	-	-	15:26	9139
15:59	16:35	-	-	16:45	17:03	9145
17:59	18:35	-	18:33	-	19:03	9153
19:59	20:35	-	-	20:45	21:09	9161

BRUSSELS Midi/Zuid	LILLE Europe	CALAIS Fréthun	ASHFORD International	EBBSFLEET International	LONDON St Pancras Int	Train no.
Saturday to Sunday						
06:51	07:28	08:02	-	-	07:56	9109
07:59	-	09:02	-	-	08:56	9113
09:29	10:05	-	-	-	10:26	9119
12:20	12:56	13:32	-	13:15	13:33	9129
14:59	15:35	-	-	-	15:56	9141
16:49	17:29	18:02	17:33	-	18:06	9149
17:59	18:35	-	18:33	-	19:03	9153
18:50	19:26	20:02	-	19:45	20:03	9157
19:59	20:35	-	-	20:45	21:09	9161
20:29	21:05	-	-	21:15	21:33	9163

Data sheet: Alcohol

MEN

Approximate Blood Alcohol Percentage

Units	Body Weight in Pounds								Effect on person
	100	120	140	160	180	200	220	240	
0	0	0	0	0	0	0	0	0	Safe Driving Limit
1	4	3	3	2	2	2	2	2	Impairment Begins
2	8	6	5	5	4	4	3	3	Impairment Begins
3	11	9	8	7	6	6	5	5	Driving Skills Significantly Affected
4	15	12	11	9	8	8	7	6	
5	19	16	13	12	11	9	9	8	Legally Intoxicated
6	23	19	16	14	13	11	10	9	
7	26	22	19	16	15	13	12	11	
8	30	25	21	19	17	15	14	13	
9	34	28	24	21	19	17	15	14	
10	38	31	27	23	21	19	17	16	

Subtract 1% for each 40 minutes that elapses after drinking

Women

Approximate Blood Alcohol Percentage

Units	Body Weight in Pounds									Effect on person
	90	100	120	140	160	180	200	220	240	
0	0	0	0	0	0	0	0	0	0	Safe Driving Limit
1	5	5	4	3	3	3	2	2	2	Impairment Begins
2	10	9	8	7	6	5	5	4	4	Driving Skills Significantly Affected
3	15	14	11	11	9	8	7	6	6	
4	20	18	15	13	11	10	9	8	8	Legally Intoxicated
5	25	23	19	16	14	13	11	10	9	
6	30	27	23	19	17	15	14	12	11	
7	35	32	27	23	20	18	16	14	13	
8	40	36	30	26	23	20	18	17	15	
9	45	41	34	29	26	23	20	19	17	
10	51	45	38	32	28	25	23	21	19	

Subtract 1% for each 40 minutes that elapses after drinking

Data sheet: Follow that car

Vehicle registration data

Vehicle Number Plates

For more information go to: **www.direct.gov.uk/motoring**

INF104

New Style Number Plates

A new number plate format was introduced on 1 September 2001.

At the same time new regulations took effect governing design, manufacture and display of number plates. The regulations introduce a standard typeface, making it easier for number plates to be read. The change applies to the plates of vehicles registered from 1 September 2001 and to all replacement plates issued from that date. The regulations **end** the use of italic, multiple stroke and other difficult to read lettering on number plates.

The Mandatory Typeface

An example of the new character typeface is shown below.

123456789
ABCDEFGH
JKLMNOPQ
RSTUVWXYZ

N.B: I and Q are not used in the new format. Q appears on some old format number plates, these "Q marks" will continue to be issued. See booklet INF46 "Registration numbers and you" for information.

Z is only used in the random element of the new format.

The Euro Plate

There is also provision, on a voluntary basis, for the display of the Euro symbol and GB national identifier on the plate. This enables motorists to dispense with a separate GB sticker when travelling within the EU, if they wish.

A New Format

A new registration mark system began on 1 September 2001 for all new vehicles being registered. The new format is as follows:

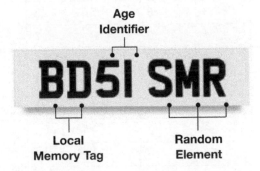

The new format will comprise seven characters and include local and age identifiers, as shown above. The local memory tag is linked to the DVLA local office where the vehicle was first registered and is intended to help witnesses recall details of the number plate.

The age identifier changes every six months in March and September. In the example above BD signifies Birmingham, 51 represents September 2001 and SMR is the random element. Lists of local memory tags and age identifiers are shown overleaf. Please note, DVLA local office identifiers may on occasion be subject to change.

Personalised Registrations

DVLA Personalised Registrations gives motorists the opportunity to purchase both prefix and current style registrations. To check the availability and price of all DVLA's Personalised Registrations visit **www.dvlaregistrations.co.uk**.

CUSTOMER SERVICE EXCELLENCE

INVESTOR IN PEOPLE

7/06

An executive agency of the Department for **Transport**

Local Memory Tags

Letter	Local Office	DVLA local office Identifier
A Anglia	Peterborough	AA AB AC AD AE AF AG AH AJ AK AL AM AN
	Norwich	AO AP AR AS AT AU
	Ipswich	AV AW AX AY
B Birmingham	Birmingham	BA – BY
C Cymru	Cardiff	CA CB CC CD CE CF CG CH CJ CK CL CM CN CO
	Swansea	CP CR CS CT CU CV
	Bangor	CW CX CY
D Deeside to Shrewsbury	Chester	DA DB DC DD DE DF DG DH DJ DK
	Shrewsbury	DL DM DN DO DP DR DS DT DU DV DW DX DY
E Essex	Chelmsford	EA – EY
F Forest & Fens	Nottingham	FA FB FC FD FE FF FG FH FJ FK FL FM FN FP
	Lincoln	FR FS FT FV FW FX FY
G Garden of England	Maidstone	GA GB GC GD GE GF GG GH GJ GK GL GM GN GO
	Brighton	GP GR GS GT GU GV GW GX GY
H Hampshire & Dorset	Bournemouth	HA HB HC HD HE HF HG HH HJ
	Portsmouth	HK HL HM HN HO HP HR HS HT HU HV HW HX HY
		(HW will be used exclusively for Isle of Wight residents)
K	Luton	KA KB KC KD KE KF KG KH KJ KK KL
	Northampton	KM KN KO KP KR KS KT KU KV KW KX KY
L London	Wimbledon	LA LB LC LD LE LF LG LH LJ
	Stanmore	LK LL LM LN LO LP LR LS LT
	Sidcup	LU LV LW LX LY
M Manchester & Merseyside	Manchester	MA – MY
		(MN + MAN Reserved for the Isle of Man)
N North	Newcastle	NA NB NC ND NE NG NH NJ NK NL NM NN NO
	Stockton	NP NR NS NT NU NV NW NX NY
O Oxford	Oxford	OA – OY
P Preston	Preston	PA PB PC PD PE PF PG PH PJ PK PL PM PN PO PP PR PS PT
	Carlisle	PU PV PW PX PY
R Reading	Reading	RA – RY
S Scotland	Glasgow	SA SB SC SD SE SF SG SH SJ
	Edinburgh	SK SL SM SN SO
	Dundee	SP SR SS ST
	Aberdeen	SU SV SW
	Inverness	SX SY
V Severn Valley	Worcester	VA – VY
W West of England	Exeter	WA WB WC WD WE WF WG WH WJ
	Truro	WK WL
	Bristol	WM WN WO WP WR WS WT WU WV WW WX WY
Y Yorkshire	Leeds	YA YB YC YD YE YF YG YH YJ YK
	Sheffield	YL YM YN YO YP YR YS YT YU
	Beverley	YV YW YX YY

N.B: DVLA cannot guarantee that any specific local memory tag or DVLA local office Identifier will be issued.

Age Identifiers

Date	Code	Date	Code
		Sept 2001 – Feb 2002	51
March 2002 – Aug 2002	02	Sept 2002 – Feb 2003	52
March 2003 – Aug 2003	03	Sept 2003 – Feb 2004	53
March 2004 – Aug 2004	04	Sept 2004 – Feb 2005	54
March 2005 – Aug 2005	05	Sept 2005 – Feb 2006	55
March 2006 – Aug 2006	06	Sept 2006 – Feb 2007	56
March 2007 – Aug 2007	07	Sept 2007 – Feb 2008	57
March 2008 – Aug 2008	08	Sept 2008 – Feb 2009	58
March 2009 – Aug 2009	09	Sept 2009 – Feb 2010	59
March 2010 – Aug 2010	10	Sept 2010 – Feb 2011	60
March 2011 – Aug 2011	11	Sept 2011 – Feb 2012	61

This pattern will continue until all permutations are exhausted.

Year indicators for car number plates issued from 1963 to 2001

Suffix series 1963–82 (letters)

Letter	Dates of issue
A	January 1963–December 1963
B	January 1964–December 1964
C	January 1965–December 1965
D	January 1966–December 1966
E	January 1967–July 1967
F	August 1967–July 1968
G	August 1968–July 1969
H	August 1969–July 1970
J	August 1970–July 1971
K	August 1971–July 1972
L	August 1972–July 1973
M	August 1973–July 1974
N	August 1974–July 1975
P	August 1975–July 1976
R	August 1976–July 1977
S	August 1977–July 1978
T	August 1978–July 1979
V	August 1979–July 1980
W	August 1980–July 1981
X	August 1981–July 1982
Y	August 1982–July 1983

Prefix series 1983–2001 (letters)

Letter	Dates of issue
A	August 1983–July 1984
B	August 1984–July 1985
C	August 1985–July 1986
D	August 1986–July 1987
E	August 1987–July 1988
F	August 1988–July 1989
G	August 1989–July 1990
H	August 1990–July 1991
J	August 1991–July 1992
K	August 1992–July 1993
L	August 1993–July 1994
M	August 1994–July 1995
N	August 1995–July 1996
P	August 1996–July 1997
R	August 1997–July 1998
S	August 1998–February 1999
T	March 1999–August 1999
V	September 1999–February 2000
W	March 2000–August 2000
X	September 2000–February 2001
Y	March 2001–August 2001

Data sheet: Bike race

Men's 1Km Time Trial / 1Km Contre la montre hommes
Final / Finale
Analysis / Analyse
Sun 27 Mar 2011

No 243 - AMRAN Muhd Arfy Qhairant (MAS)

Distance	Time	Rank	Lap Time
125m	12.165	20	
250m	19.645	20	19.645
375m	27.042	20	
500m	34.588	20	14.943
625m	42.478	20	
750m	50.702	20	16.114
875m	59.373	20	
1000m	1:08.450	20	17.748

No 284 - KUCZYNSKI Kamil (POL)

Distance	Time	Rank	Lap Time
125m	11.712	10	
250m	18.873	10	18.873
375m	25.805	9	
500m	32.840	9	13.967
625m	40.089	11	
750m	47.630	12	14.790
875m	55.518	12	
1000m	1:03.791	13	16.161

No 297 - KUBEEV Andrey (RUS)

Distance	Time	Rank	Lap Time
125m	12.031	18	
250m	19.222	16	19.222
375m	26.072	15	
500m	32.912	13	13.690
625m	39.924	9	
750m	47.186	9	14.274
875m	54.824	9	
1000m	1:02.838	8	15.652

No 119 - FABIAN HERNANDO Puerta Zapata (COL)

Distance	Time	Rank	Lap Time
125m	12.025	17	
250m	19.171	14	19.171
375m	26.024	14	
500m	32.898	11	13.727
625m	39.981	10	
750m	47.444	10	14.546
875m	55.318	10	
1000m	1:03.653	12	16.209

No 322 - BOLIBRUKH Yevhen (UKR)

Distance	Time	Rank	Lap Time
125m	11.746	12	
250m	18.998	12	18.998
375m	25.929	12	
500m	32.907	12	13.909
625m	40.133	12	
750m	47.651	13	14.744
875m	55.568	13	
1000m	1:03.846	14	16.195

No 342 - PULGAR ARAUJO Angel Ramiro (VEN)

Distance	Time	Rank	Lap Time
125m	11.374	4	
250m	18.663	7	18.663
375m	25.908	11	
500m	33.250	16	14.587
625m	40.797	17	
750m	48.576	18	15.326
875m	56.639	18	
1000m	1:05.083	18	16.507

No 187 - EILERS Joachim (GER)

Distance	Time	Rank	Lap Time
125m	11.349	3	
250m	18.320	4	18.320
375m	25.135	3	
500m	31.977	5	13.657
625m	39.052	5	
750m	46.402	5	14.425
875m	54.138	5	
1000m	1:02.296	5	15.894

No 246 - TISIN Mohd Rizal (MAS)

Distance	Time	Rank	Lap Time
125m	11.971	16	
250m	19.186	15	19.186
375m	26.075	16	
500m	33.007	15	13.821
625m	40.162	14	
750m	47.579	11	14.572
875m	55.416	11	
1000m	1:03.651	11	16.072

Men's 1Km Time Trial / 1Km Contre la montre hommes
Final / Finale
Analysis / Analyse
Sun 27 Mar 2011

No 218 - CECI Francesco (ITA)

Distance	Time	Rank	Lap Time
125m	12.116	19	
250m	19.485	19	19.485
375m	26.590	19	
500m	33.697	19	14.212
625m	41.112	19	
750m	48.792	19	15.095
875m	56.825	19	
1000m	1:05.193	19	16.401

No 230 - NITTA Yudai (JPN)

Distance	Time	Rank	Lap Time
125m	11.707	9	
250m	18.848	9	18.848
375m	25.822	10	
500m	32.881	10	14.033
625m	40.179	15	
750m	47.782	15	14.901
875m	55.790	15	
1000m	1:04.201	16	16.419

No 277 - VAN VELTHOOVEN Simon (NZL)

Distance	Time	Rank	Lap Time
125m			
250m			
375m			
500m			
625m			
750m			
875m			
1000m			

No 304 - ZHURKIN Nikolay (RUS)

Distance	Time	Rank	Lap Time
125m	11.912	14	
250m	19.114	13	19.114
375m	25.982	13	
500m	32.926	14	13.812
625m	40.135	13	
750m	47.719	14	14.793
875m	55.811	16	
1000m	1:04.312	17	16.593

No 159 - LAFARGUE Quentin (FRA)

Distance	Time	Rank	Lap Time
125m	11.679	8	
250m	18.839	8	18.839
375m	25.622	7	
500m	32.447	7	13.608
625m	39.500	7	
750m	46.840	6	14.393
875m	54.522	6	
1000m	1:02.582	6	15.742

No 252 - HAAK Hugo (NED)

Distance	Time	Rank	Lap Time
125m	11.196	2	
250m	18.245	2	18.245
375m	25.165	5	
500m	32.228	6	13.983
625m	39.470	6	
750m	46.944	7	14.716
875m	54.763	8	
1000m	1:02.897	9	15.953

No 286 - TEKLINSKI Adrian (POL)

Distance	Time	Rank	Lap Time
125m	11.910	13	
250m	19.333	17	19.333
375m	26.502	18	
500m	33.666	18	14.333
625m	40.940	18	
750m	48.407	17	14.741
875m	56.101	17	
1000m	1:04.086	15	15.679

No 129 - BABEK Tomas (CZE)

Distance	Time	Rank	Lap Time
125m	11.718	11	
250m	18.882	11	18.882
375m	25.707	8	
500m	32.546	8	13.664
625m	39.617	8	
750m	46.983	8	14.437
875m	54.729	7	
1000m	1:02.788	7	15.805

Men's 1Km Time Trial / 1Km Contre la montre hommes
Final / Finale
Analysis / Analyse
Sun 27 Mar 2011

No 270 - DAWKINS Edward (NZL)

Distance	Time	Rank	Lap Time
125m	11.945	15	
250m	19.357	18	19.357
375m	26.405	17	
500m	33.424	17	14.067
625m	40.588	16	
750m	47.941	16	14.517
875m	55.571	14	
1000m	1:03.534	10	15.593

No 192 - NIMKE Stefan (GER)

Distance	Time	Rank	Lap Time
125m	11.444	6	
250m	18.412	5	18.412
375m	25.137	4	
500m	31.888	3	13.476
625m	38.794	3	
750m	45.871	3	13.983
875m	53.178	2	
1000m	1:00.793	1	14.922

No 162 - PERVIS François (FRA)

Distance	Time	Rank	Lap Time
125m	11.401	5	
250m	18.300	3	18.300
375m	24.938	2	
500m	31.618	2	13.318
625m	38.524	2	
750m	45.723	2	14.105
875m	53.284	3	
1000m	1:01.228	3	15.505

No 156 - D'ALMEIDA Michaël (FRA)

Distance	Time	Rank	Lap Time
125m	11.523	7	
250m	18.515	6	18.515
375m	25.233	6	
500m	31.929	4	13.414
625m	38.832	4	
750m	46.004	4	14.075
875m	53.575	4	
1000m	1:01.481	4	15.477

No 255 - MULDER Teun (NED)

Distance	Time	Rank	Lap Time
125m	11.100	1	
250m	18.056	1	18.056
375m	24.714	1	
500m	31.411	1	13.355
625m	38.324	1	
750m	45.542	1	14.131
875m	53.151	1	
1000m	1:01.179	2	15.637

Data sheet: Give us a job

All these jobs were advertised in Glasgow on one day in 2009.

Freight sales coordinator £14 000 pa	Payroll officer £15 000 pa	Occupational therapist £25 per hour	Care worker £9 per hour
Technical sales engineer £30 000 pa	Sales executive £22 000 pa	Accountant £29 000 pa	Catering assistant £5.75 per hour
Conference assistant £6 per hour	Ledger clerk £18 000 pa	Asbestos analyst £19 000 pa	Nursery class teacher £120 per day
Beauty clinic manager £32 000 pa	Accounts clerk £7.60 per hour	Ultrasound operator £27 000 pa	Financial adviser £18 400 pa
Security adviser £20 000 pa	Engineering performance assessor £21 000 pa	Mortgage adviser £24 000 pa	Recruitment consultant £17 500 pa
Hardware sales executive £19 400 pa	Personal injury paralegal £20 400 pa	Physiotherapist £22 per hour	Human resources adviser £26 000 pa
Social care worker £14 700 pa	Investigator (anti-money laundering) £8.50 per hour	Teacher of mathematics £140 per day	CCTV engineer £24 000 pa
Events and sales executive £16 200 pa	Insurance broker £18 600 pa	Speech therapist £21 per hour	Maintenance technician £26 500 pa
Bank customer adviser £12 400 pa	Assistant financial analyst £9.50 per hour	Office manager £20 500 pa	Motor claims handler £16 500 pa
Packaging designer £16 per hour	Sous chef in a restaurant £23 000 pa	Call centre operator £14 500 pa	Nursery manager £19 500 pa
Fundraising coordinator £20 500 pa	Personal trainer £17 000 pa	Telemarketer researcher £13 000 pa	Maintenance engineer £26 500 pa
Social worker £21 per hour	Teaching assistant £10 per hour	Secretary £7.50 per hour	Assistant restaurant manager £18 500 pa

Data sheet: Populations

Country	Population	Area(sq km)
Afghanistan	33 609 937	647 500
Akrotiri	15 700	123
Albania	3 639 453	28 748
Algeria	34 178 188	2 381 740
American Samoa	65 628	199
Andorra	83 888	468
Angola	12 799 293	1 246 700
Anguilla	14 436	102
Antigua and Barbuda	85 632	443
Argentina	40 913 584	2 766 890
Armenia	2 967 004	29 743
Aruba	103 065	193
Australia	21 262 641	7 686 850
Austria	8 210 281	83 870
Azerbaijan	8 238 672	86 600
Bahamas The	309 156	13 940
Bahrain	727 785	665
Bangladesh	156 050 883	144 000
Barbados	284 589	431
Belarus	9 648 533	207 600
Belgium	10 414 336	30 528
Belize	307 899	22 966
Benin	8 791 832	112 620
Bermuda	67 837	53
Bhutan	691 141	47 000
Bolivia	9 775 246	1 098 580
Bosnia and Herzegovina	4 613 414	51 209
Botswana	1 990 876	600 370
Brazil	198 739 269	8 511 965
British Virgin Islands	24 491	153
Brunei	388 190	5 770
Bulgaria	7 204 687	110 910
Burkina Faso	15 746 232	274 200
Burma	48 137 741	678 500
Burundi	8 988 091	27 830
Cambodia	14 494 293	181 040
Cameroon	18 879 301	475 440
Canada	33 487 208	9 984 670
Cape Verde	429 474	4 033
Cayman Islands	49 035	262
Central African Republic	4 511 488	622 984
Chad	10 329 208	1 284 000
Chile	16 601 707	756 950
China	1 338 612 968	9 596 960
Christmas Island	1 402	135
Cocos (Keeling) Islands	596	14
Colombia	45 644 023	1 138 910
Comoros	752 438	2 170
Congo, Democratic Republic of the	68 692 542	2 345 410
Congo, Republic of the	4 012 809	342 000
Cook Islands	11 870	237
Costa Rica	4 253 877	51 100
Cote d'Ivoire	20 617 068	322 460
Croatia	4 489 409	56 542
Cuba	11 451 652	110 860
Cyprus	796 740	9 250
Czech Republic	10 211 904	78 866
Denmark	5 500 510	43 094

Country	Population	Area(sq km)
Dhekelia	15 700	131
Djibouti	516 055	23 000
Dominica	72 660	754
Dominican Republic	9 650 054	48 730
Ecuador	14 573 101	283 560
Egypt	83 082 869	1 001 450
El Salvador	7 185 218	21 040
Equatorial Guinea	633 441	28 051
Eritrea	5 647 168	121 320
Estonia	1 299 371	45 226
Ethiopia	85 237 338	1 127 127
European Union	491 582 852	4 324 782
Falkland Islands	3 140	12 173
Faroe Islands	48 856	1 399
Fiji	944 720	18 270
Finland	5 250 275	338 145
France	64 057 792	643 427
French Polynesia	287 032	4 167
Gabon	1 514 993	267 667
Gambia The	1 782 893	11 300
Gaza Strip	1 551 859	360
Georgia	4 615 807	69 700
Germany	82 329 758	357 021
Ghana	23 832 495	239 460
Gibraltar	28 034	7
Greece	10 737 428	131 940
Greenland	57 600	2 166 086
Grenada	90 739	344
Guam	178 430	541
Guatemala	13 276 517	108 890
Guernsey	65 870	78
Guinea	10 057 975	245 857
Guinea-Bissau	1 533 964	36 120
Guyana	772 298	214 970
Haiti	9 035 536	27 750
Holy See (Vatican City)	826	0.4
Honduras	7 792 854	112 090
Hong Kong	7 055 071	1 092
Hungary	9 905 596	93 030
Iceland	306 694	103 000
India	1 166 079 217	3 287 590
Indonesia	240 271 522	1 919 440
Iran	66 429 284	1 648 000
Iraq	28 945 657	437 072
Ireland	4 203 200	70 280
Isle of Man	76 512	572
Israel	7 233 701	20 770
Italy	58 126 212	301 230
Jamaica	2 825 928	10 991
Japan	127 078 679	377 835
Jersey	91 626	116
Jordan	6 342 948	92 300
Kazakhstan	15 399 437	2 717 300
Kenya	39 002 772	582 650
Kiribati	112 850	811
Korea North	22 665 345	120 540
Korea South	48 508 972	98 480
Kosovo	1 804 838	10 887
Kuwait	2 691 158	17 820
Kyrgyzstan	5 431 747	198 500
Laos	6 834 942	236 800

Problem-solving Skills for Maths

Country	Population	Area(sq km)
Latvia	2 231 503	64 589
Lebanon	4 017 095	10 400
Lesotho	2 130 819	30 355
Liberia	3 441 790	111 370
Libya	6 310 434	1 759 540
Liechtenstein	34 761	160
Lithuania	3 555 179	65 300
Luxembourg	491 775	2 586
Macau	559 846	28
Macedonia	2 066 718	25 333
Madagascar	20 653 556	587 040
Malawi	14 268 711	118 480
Malaysia	25 715 819	329 750
Maldives	396 334	300
Mali	12 666 987	1 240 000
Malta	405 165	316
Marshall Islands	64 522	181
Mauritania	3 129 486	1 030 700
Mauritius	1 284 264	2 040
Mayotte	223 765	374
Mexico	111 211 789	1 972 550
Micronesia, Federated States of	107 434	702
Moldova	4 320 748	33 843
Monaco	32 965	2
Mongolia	3 041 142	1 564 116
Montenegro	672 180	14 026
Montserrat	5 097	102
Morocco	34 859 364	446 550
Mozambique	21 669 278	801 590
Namibia	2 108 665	825 418
Nauru	14 019	21
Nepal	28 563 377	147 181
Netherlands	16 715 999	41 526
Netherlands Antilles	227 049	960
New Caledonia	227 436	19 060
New Zealand	4 213 418	268 680
Nicaragua	5 891 199	129 494
Niger	15 306 252	1 267 000
Nigeria	149 229 090	923 768
Niue	1 398	260
Norfolk Island	2 141	35
Northern Mariana Islands	88 662	477
Norway	4 660 539	323 802
Oman	3 418 085	212 460
Pakistan	176 242 949	803 940
Palau	20 796	458
Panama	3 360 474	78 200
Papua New Guinea	6 057 263	462 840
Paraguay	6 995 655	406 750
Peru	29 546 963	1 285 220
Philippines	97 976 603	300 000
Pitcairn Islands	48	47
Poland	38 482 919	312 679
Portugal	10 707 924	92 391
Puerto Rico	3 971 020	13 790
Qatar	833 285	11 437
Romania	22 215 421	237 500
Russia	140 041 247	17 075 200
Rwanda	10 473 282	26 338
Saint Barthelemy	7 448	21
Saint Helena	7 637	413
Saint Kitts and Nevis	40 131	261

137

Country	Population	Area(sq km)
Saint Lucia	160 267	616
Saint Martin	29 820	54
Saint Pierre and Miquelon	7 051	242
Saint Vincent and the Grenadines	104 574	389
Samoa	219 998	2 944
San Marino	30 324	61
Sao Tome and Principe	212 679	1 001
Saudi Arabia	28 686 633	2 149 690
Senegal	13 711 597	196 190
Serbia	10 159 046	77 474
Seychelles	87 476	455
Sierra Leone	6 440 053	71 740
Singapore	4 657 542	693
Slovakia	5 463 046	48 845
Slovenia	2 005 692	20 273
Solomon Islands	595 613	28 450
Somalia	9 832 017	637 657
South Africa	49 052 489	1 219 912
Spain	40 525 002	504 782
Sri Lanka	21 324 791	65 610
Sudan	41 087 825	2 505 810
Suriname	481 267	163 270
Svalbard	2 116	61 020
Swaziland	1 123 913	17 363
Sweden	9 059 651	449 964
Switzerland	7 604 467	41 290
Syria	20 178 485	185 180
Taiwan	22 974 347	35 980
Tajikistan	7 349 145	143 100
Tanzania	41 048 532	945 087
Thailand	65 905 410	514 000
Timor-Leste	1 131 612	15 007
Togo	6 019 877	56 785
Tokelau	1 416	10
Tonga	120 898	748
Trinidad and Tobago	1 229 953	5 128
Tunisia	10 486 339	163 610
Turkey	76 805 524	780 580
Turkmenistan	4 884 887	488 100
Turks and Caicos Islands	22 942	430
Tuvalu	12 373	26
Uganda	32 369 558	236 040
Ukraine	45 700 395	603 700
United Arab Emirates	4 798 491	83 600
United Kingdom	61 113 205	244 820
United States	307 212 123	9 826 630
Uruguay	3 494 382	176 220
Uzbekistan	27 606 007	447 400
Vanuatu	218 519	12 200
Venezuela	26 814 843	912 050
Vietnam	86 967 524	329 560
Virgin Islands	109 825	1 910
Wallis and Futuna	15 289	274
West Bank	2 461 267	5 860
Western Sahara	405 210	266 000
Yemen	23 822 783	527 970
Zambia	11 862 740	752 614
Zimbabwe	11 392 629	390 580
World	6 706 993 152	510 072 000

Source: www.cia.gov April 2009

Answers

Endangered species

Warm-up questions
1 20–40 years **2** Longer **3** Approximately 42 000
4 Heavier **5** Male

Task 3
1 Up to 33 metres **2** 182 kg **3** 2% **4** 2 grams
5 25 000 **6** 1.6%

Football

Task 1
1 180 **2** +36 **3** 23 **4** Overall there must be the
same number of goals scored for as goals scored against

5 15 **6** 108 **7 a** 30 **b** 20 wins, 10 draws
8 14 draws, 5 losses **9** 4 wins, 8 draws
10 a True **b** False **c** False **d** False

Task 2
1 32 **2** 1024 **3** 2^n

Task 3
The number of possible half-time scores for a final
score of $h - a$ is $(h + 1)(a + 1)$

Paving

Warm-up questions
1 100 **2** 120 cm **3** £258
4 £258 + £378 = £636 **5** 600 m²
6 £11 **7** 25 **8** 20 **9** 5 with £10 left over
10 20 of each

Money matters 1: Pay

Task 1
Aftab: £252; Betty: £243.20; Colin: £513;
Dierdre: £110.40; Eddy: 35 hours; Frank: 40 hours;
Gus: £2.75 per hour; Hinna: £9.80 per hour;
Ian: 18–20 years old; Jemima: 16 or 17 years old

Task 2
Alf, £410; Belinda, £288.80; Chas, £552;
Dave £165.60; Edith, 5 hours; Francis, 4 hours;
Gaynor, £3.20 per hour; Henry, £9 per hour;
Iris, 18–20 years old; Jack, 16 or 17 years old.

Task 3
Pete: £30 000; Quinlan: £40 200; Rosie: £1875;
Sue: £6210; Teresa: £50 160

Extension questions
1 £598.50 **2** £290.09 **3** £7.50
4 18–20 years old **5** £225.60

Scotland

Task 1
1 13% **2** Industrial west **3** 10 000 km
4 1340 m **5** 74 **6** 43 **7** 60.1°C
8 70% **9** 5672 km² **10** 887 400

Task 2

Letter	Tally	Frequency
A	IIII I	7
E	IIII	5
I	IIII	5
B	III	3
C	IIII	4
N	IIII I	7
others	IIII IIII IIII IIII I	21

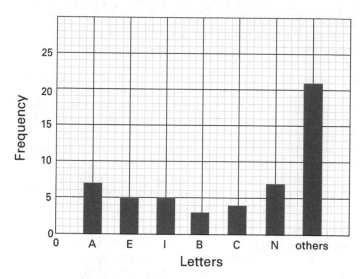

3 a 4/52 **b** 48/52

Task 3
1 Edinburgh is east of Glasgow
2 Perth is north-east of Stirling
3 Dundee is south-west of Aberdeen
4 Inverness is north of Glasgow
5 Dundee is south of Wick
6 Stranraer is south-west of Dunfermline

Bricklaying patterns

Task 1
Flemish bond has
- a vertical line of symmetry through the centre of any brick
- a horizontal line of symmetry through the centre of any line of bricks
- rotational symmetry of order 2 through the centre of any brick
- rotational symmetry of order 2 through a point halfway between the base of one header and the top of the adjacent header on the row below.

Task 2
Stack bond has:
- vertical lines of symmetry through the joins or the centres of any bricks
- horizontal lines of symmetry through the joins or the centres of any bricks
- rotational symmetry of order 2 through the centres of bricks and the centres of any edge.

English bond has:
- vertical line symmetry through the centre of any brick
- horizontal line symmetry through the centre of any row of bricks
- rotational symmetry of order 2 through the centre of any brick or the centre of the vertical edge of any stretcher.

English cross bond has:
- vertical line symmetry through the centre of any brick
- horizontal line symmetry through the centre of any row of bricks
- rotational symmetry of order 2 through the centre of any stretcher and the centre of any vertical join of 2 headers.

Monk bond has:
- vertical line symmetry through the centre of any header and the vertical join of any two stretchers
- horizontal line symmetry through the centre of any row of bricks
- rotational symmetry of order 2 through the centre of any header and the centre of any vertical join of 2 stretchers.

Answers

Task 3
Herringbone has rotational symmetry of order 2 through the centre of the join where two vertical or horizontal bricks touch.

Basketweave has vertical and horizontal line symmetry through the join of any two adjacent horizontal or vertical bricks and rotational symmetry of order 2 about the centre of the join of any two adjacent horizontal or vertical bricks.

Pinwheel has rotational symmetry of order 4 through the centre of the small centre square.

De laRobia weave has vertical and horizontal symmetry through the centre of any brick and rotational symmetry of order 2 through the centre of any brick.

Deliveries
Warm-up questions
1 23 miles   2 24 minutes   3 6 journeys
4 $150 \times 80p = £120$ 5 $6 \times £120 = £720$

Task 1
32 miles

Task 2
46 miles

Task 3
$10 + 12 + 8 = 30$ miles and $6 + 11 + 12 = 29$ miles. Therefore it is more expensive to use two vans but deliveries are quicker

Task 4
With one van it is 618 miles from Glenrothes to Inverness to Paisley to Stranraer to Hawick and back to Glenrothes.

With two vans there are various answers. It would be best to send one van to Inverness and the other van to Paisley, Stranraer and Hawick.

Water
Task 1
1 Toilets: 38 litres; baths: 50 litres; drinking/cooking: 20 litres; dishwashing: 12 litres; clothes: 21 litres; other: 11 litres

2 4.55

Task 2
Volume of a brick $= 1432$ cm^3 $= 1.432$ litres,
6 flushes $\times 365 = 2190$ flushes;
saving over a year: $2190 \times 1.432 = 3136 \approx 3140$ litres

Task 3
1 £306.11   2 £655.95
3 $£568.49 - £349.84 = £218.65$

Task 4
1 $3 \times 150 \times 365 = 164\,250$ litres,
$164\,250 \times 0.5p = £821.25$
2 Council tax band F pays £568.49, so the water meter is more expensive by $£821.25 - £568.49 = £252.76$

Task 5
1 77.6 litres
2 $4 \times 365 \times 140 = 204\,400$ litres; this is $204\,400 \div 77.6 = 2634$ bathtubs

Safe flying over the UK
1 FL 65   2 28000 feet   3 Flight levels go up in fives (500 feet at a time) so FL30 is followed by FL35.
4 FL 110 or FL 130   5 FL 100 or FL 120
6 FL 100 or FL 120   7 FL 110 or FL 130
8 1000 feet   9 2000 feet   10 4000 feet

Darts
1 a 69 b 91 c 54 d 77 e 83
2 a Any finish that gives a total of 60 for the first two darts, e.g. D20, S20
   b Any finish that gives a total of 50 for the first two darts, e.g. D20, S10
   c Any finish that gives a total of 68 for the first two darts, e.g. T20, S8
3 a 180 b 23 4 a e.g. T19, D9 b S60, D19 c S1, D1
5 The largest number that can be scored with one dart is 60, so $99 - 60 = 39$. 39 cannot be scored with 1 dart.
6 23, 29, 31, 35, 37, 41, 43, 44, 46, 47, 49, 52, 53, 55, 56, 58, 59
7 a 10 and 14; 9 and 15   b $18 + 7 = 25$
   c $12 + 2 = 14$
8 a 16 and 4; 5 and 17   b $14 - 10 = 4$
   c $19 - 1 = 18$
9 a 10 and 15, 11 and 14   b $6 + 10 = 16$
   c $1 + 18$, $13 + 6$, $2 + 17$, $8 + 11$ d $19 + 7 = 26$
10 Adding pairs of numbers $10 + 11$, $9 + 12$, $8 + 13$, $7 + 14$, $6 + 15$, etc. all give a total of 21.
There are 10 lots of these, so $10 \times 21 = 210$

Extension question
First three darts: $3 \times T20 = 180$, leaving
$501 - 180 = 321$; second three darts: $3 + T20 = 180$,
leaving $321 - 180 = 141$; 141 has several three-dart finishes, e.g. T20, T19, D12 or T19, T18, D15

Bridges
Task 1
1 6 years   2 835000   3 125000 cubic metres
4 Askøy, Severn, Ponte 25 de Abril   5 49.6 km
6 2 years   7 398 m   8 £20.20   9 82.56p Most of the vehicles were cars, and there were slightly more buses and lorries than there were motor cycles.
10 30%

Task 2
Humber Bridge. At 1410 m, this is the only bridge in the table with a span 40% Longer than the Forth Road Bridge.

Money matters 2
1 Annual income: £11440; taxable income: £3965; 20% of £3965 = £793; weekly tax: **£15.25**; NI: 12% of £220 – £139 = **£9.72**; total deductions: **£24.97**; weekly take-home pay: **£195.03**

2 Taxable income: £36525.; tax paid: 20% of £35000 + 40% of £1525 = £7610; monthly tax: **£634.17**; weekly pay: £846.15; NI: 12% of £678 + 2% of £29.15 = £81.94; monthly NI: £81.94 × 52 ÷ 12 = **£355.07**; monthly net salary: £3666.67 – (£355.07 + £634.17) = **£2677.43**

3 Annual salary: £30000; taxable income £22525; 20% of £22525 = £4505; monthly tax: **£375.42**; weekly income: £30000 ÷ 52 = £576.92. NI: 12% of £437.92 = £52.55; monthly NI: **£227.72**; monthly net salary: £2500 – £375.42 – £227.72 = **£1896.86**

4 Annual income: £28600; taxable income: £21125; 20% of £21125 = £4225; weekly tax: £4225 ÷ 52 = **£81.25**; NI: 12% of £411 = **£49.32**; net weekly wage: £550 – £81.25 – £49.32 = **£419.43**

5 Mr T: annual income: £19 760; taxable income: £12 285; 20% of £12 285 = £2457; weekly tax: **£47.25**; NI: 12% of £241 = **£28.92**; Mrs T: annual income: £6760; taxable income: 0; weekly tax: 0; NI: 12% of 0 = 0; total weekly income: £510; total weekly deductions: (£47.25 + £28.92); net weekly income: **£433.83**

Money matters 2

6 Mr U: taxable income: £44 525; tax: 20% of £35 000 + 40% of £9525 = £10 810; monthly tax: **£900.83**; weekly income: £1000; N1: 12% of £678 + 2% of £183 = £85.02; monthly NI: **£368.42**; Ms V: taxable income: £50 525. Tax: 20% of £35 000 + 40% of £15 525 = £13 210; monthly tax: **£1100.83**; weekly income: £1115.38; N1: 12% of £678 + 2% of £298.38 + £87.33; monthly NI: **£378.43**; monthly joint income: £9166.67; monthly deductions: £900.83 + £368.42 + £1100.83 + £378.43; net monthly joint income: **£6418.16**

7 Before pay rise: annual income: £36 000; taxable income: £28 525; 20% of £28 525 = £5705; monthly tax: **£475.42**; weekly pay: £692.31; 12% of £553.31 = £66.40; monthly NI: **£287.72**; take-home monthly before rise: **£2236.86**.

After pay rise: annual income: 6 × £3000 + 6 × £3500 = £39 000; taxable income: £31 525; 20% of £31 525 = £6305; monthly tax: **£525.42**; weekly pay: £807.69; 12% of £668.69 = £80.24; monthly NI: **£347.71**; take-home monthly pay after rise: **£2626.87**; increase: **£390.01**.

8 a Taxable income = £16 525, tax = 20% of £16 525 = £3305

b Taxable income = £8525, tax = 20% of £8525 = £1705

c Tax paid for 8 months = £3305 ÷ 12 × 8 = £2203.33. Tax rebate = £2203.33 − £1705 = **£498.33**

At the gym

Task 1
Most people will use the nearest shower; as shower 1 is nearest the entrance this will be used first if it is available. That is why shower 1 has almost run out and shower 5 hasn't been used.

Task 2
Some men come in very early and then a few more arrive before work. It then gets quieter until lunchtime when more men arrive; it goes quiet during the afternoon and peaks after work.

At the gym (continued)

Task 3
Between 6 am and 9 am 60 men visit the gym. In that time 1 litre of soap is used. This is 1000 ÷ 60 = 16.67 ml per person on average.

During the day 320 men visit the gym. This is 320 × 16.67 = 5334.4 ml of soap ≈ 5.3 litres.

Task 4
The bar chart should run from 9 am to 5 pm, with most bars about 20 units high but with the bars around 12 midday to 2 pm slightly higher.

Task 5
a Approximately 1.1 litres

b On the previous Monday, in shower 1, 450 ml of soap had been used by 9 am. This Monday 300 ml has been used.

This is about 100 ml per hour, so the soap in shower 1 should last 5 hours. The latest time to get them checked is 11 am.

Task 6 (extension)
Any sensible program, for example:

15 minutes kick boxing (150 calories), 10 minutes light weights (70 calories), 10 minutes light rowing (70 calories), 10 minutes jogging on treadmill (100 calories), 15 minutes aerobics (105 calories). Total 1 hour and 495 calories.

Tour de France: 5900 calories per day on average.

Marathon: 2600 calories during a marathon.

Football match: 1000 calories for a professional player.

Money matters 3
Warm-up questions
1 a £5000 b £25 000 c £78 d £390

2 a Monthly b £80 000 c £850 d £4250

3 a 1284% b 25% c Short-term loan

4 a Weekly b £500 c Short-term loan with weekly repayments

5 a £20 000 b 5 years c Repayment insurance

Task 1 see table at foot of page

Money matters 4
Warm-up questions
1 a An investment for which the interest rate is fixed and the investment period is also fixed b No, £5000 is the minimum investment c £22.50 d £112.50

2 a On the internet b 2.50% gross AER c 0.1% gross AER

3 a £500 b 1.5% gross AER c No penalty

Money matters 3 Task 1

Name of company	Online, branch, shop or telephone	Amount that can be borrowed	Repayments: one payment or weekly or monthly	APR
TopService Personal Loans	Online	£5000 to £25 000	Monthly	7.8%
Gordon's Cheap Loans	Branch or shop	£5000 to £100 000	Monthly	8.5%
The Instant Cash Loans Company	Telephone	Up to £1000	One payment	1284%
Cash around the Corner Loans	Shop	£50 to £500	Weekly	189.2%
The Lending Bank	Online or branch	Up to £20 000	Monthly	9.9%

4 a £1000 **b** After 12 months, provided you have kept at least £1000 in the account

5 a 180 day loss of interest on the amount withdrawn during the fixed rate period **b** £1
c The amount received after basic rate tax per year (per annum)

Shuffleboard

Task 1

a Red 23, yellow 15 **b** Red 30, yellow 34 **c** Red −5, yellow 29

Task 2

All scores apart from 9, 19, 29 and 39 are possible.

Task 3

	Mean	Mode	Range	Lowest Score
Alf	22	28	27	7
Brenda	22	*	≤14, ≥6	≥16, ≤20
Clara	31.5	34	24	16

*You cannot work out Brenda's mode.

Clara should be picked because she has the highest mean score and has few low scores. You could pick either Alf or Brenda as reserve. They have the same mean score. Brenda is very consistent. Alf has more scores than Brenda greater than 30, but he also has more score less than 16.

Task 4 (extension)

Isosceles triangle: 3 square feet; two '8' trapezia, 4.5 square feet each; two '7' trapezia, 7.5 square feet each; rectangle 9 square feet

P(10) = 0.083, P(8) = 0.25, P(7) = 0.417, P(−10) = 0.25

Money matters 5

Task 1

1 £802.43 **2** £1204.28 **3** £195 120
4 £202 318.20 **5** £200.60 **6** £153.30
7 £173.34 **8** £28 251 **9** £385.04
10 £1930.74

Time zones

1 10 am **2** 4 pm **3** 20 h 30 min
4 4 hours (1 pm to 5 pm) **5** 8 pm **6** 4.30 pm
7 London: 6 pm, Los Angeles: 10 am, Moscow: 9 pm, Karachi: 11 pm
8 AU1: 12:55; AU2: 12:30; AU3: 8 h 20 m; AU4: 9 h 30 m; AU5: 07:15
9 Andrew: 12:55; Carol: 07:55 **10** 12 midday and 2 pm

Stopping distances

Task 1

1 16 **2** 15 m **3** 55 m **4** 36 m
5 96 m **6** 13.5 m **7** 84.5 m **8** 24
9 See the graph below.

10 See the graph below; thinking: 24 m; braking: about 104 m; total: about 128 m.

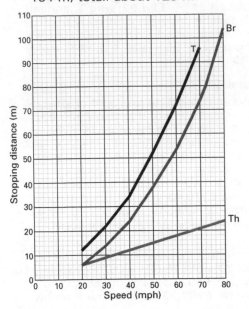

Task 2

1

Speed (mph)	30	40	50	60	70
3-second (ft)	138.5	182	220.5	265.5	310.5
6-second (ft)	267	354	441	531	621

2

Speed (mph)	30	40	50	60	70
3-second (m)	42	55	67	81	95
6-second (m)	81	108	134	162	189

3 At the lowest speed the US table advises about double the UK distance, but they get closer as the speed increases and they agree at 70 mph.

4

Speed (mph)	30	40	50	60	70
Formula (ft)	75	120	175	240	315
Formula (m)	23	37	53	73	96
Chart	23	36	53	73	96

The formula agrees with the chart to within 1 metre.

Task 3

1 120 ft **2** 2 s **3 a** 30 mph **b** 100 ft **c** 1.5 s
4 0.25 s **5** 30 mph **6** 100 ft **7** 0.25 s
8 75 ft **9** 12 mph **10** See the graph below.

Emergency braking from 60 mph

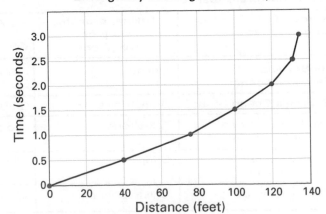

Task 4 (extension)

You should point out that the risk of a fatality is 0.25 at 30 mph, is 0.16 at 20 mph and that schools can be busy at certain times of the day with students crossing roads and not paying attention as they meet their friends.

Climate change

Task 2

1 90 years   **2** 7 years

3 434 000 000

4 0.5 metres   **5** 300 million   **6** 21.9 million

7 9/10   **8** 0.024%

Growing, growing, grown

1 a Increasing powers of 3 up to 3^9. Values are:
   1, 3, 9, 27, 81, 243, 729, 2187, 6561

b Increasing powers of 5 up to 5^9. Values are:
   1, 5, 25, 125, 625, 3125, 15 625, 78 125, 390 625

c Increasing calculations up to 3×2^9. Values are:
   3, 6, 12, 24, 48, 96, 192, 384, 768, 1536

2 0.001, 10, 1000, 10 000; 1/10 000, 1/1000, 100/1, 1000/1, 10 000/1; 10^{-4}, 10^{-2}

3 a First level at A: 32 each side; second level: 16, 32 (16 + 16), 16; third level: 8, 24 (8 + 16), 24 (16 + 8), 8; fourth level: 4, 16 (4 + 12), 24 (12 + 12), 16 (12 + 4), 4

b Tray 1: 4; tray 3: 24; tray 4: 16; tray 5: 4

4 a $2^{25} \approx 33\,500\,000$, $2^{26} \approx 67\,000\,000$, so on about 26 or 27 January.

b $2^{32} \approx 4\,300$ million, $2^{33} \approx 8\,600$ million so about 3 or 4 February. Well before St Valentine's day.

5 The number of squares builds up as square numbers to give a final total of 204.

6 The man's weight in grams is $80 \times 1000 = 80\,000$

The man's value in pounds is $80\,000 \times 30 =$ **£2 400 000**

The total number of grains is $2^{64} - 1 \approx 1.845 \times 10^{19}$

This number of grains will weigh
$1.845 \times 10^{19} \div 60\,000 = 3.075 \times 10^{14}$ kg

This weight of rice is $3.075 \times 10^{14} \div 1000 = 3.075 \times 10^{11}$ tonnes.

This weight of rice will cost $3.075 \times 10^{11} \times 400 = £1.230 \times 10^{14}$

This is approximately **£123 000 000 000 000** or 123 million million (123 trillion) pounds.

It looks like the old man got the better deal.

Task 1 (extension)

The Fibonacci sequence is formed by adding the two previous terms to get the next term. It takes 17 terms to get over 1000. The 16th term is 987, the 17th is 1597.

Dividing subsequent terms leads to 1.618…, which is called the **Golden ratio**.

In nature Fibonacci occurs in seashells, seed heads and the growth of rabbit colonies.

Venting gas appliances

Task 1

1 0   **2** 3.25 ft   **3** 18/12   **4** 2.5 feet

5

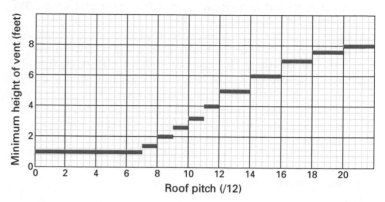

Task 2

a No. Roof pitch is 6/12, so minimum height is 1 foot but the vent pipe is only 6 feet from the vertical wall; the minimum is 8 feet.

b Yes. Pipe is over 8 feet from the vertical wall. Roof pitch is 12/12 so the minimum pipe height is 4 feet. Pipe is just over 4 feet high.

Task 3

a Pipe at least 8 feet from vertical wall. Pitch is between 9/12 and 10/12 so the vertical height must be at least 2.5 feet.

b Pipe at least 8 feet from vertical wall. Pitch is 18/12 so the vertical height must be at least 7 feet.

Task 4

1 20 feet   **2** 6 inches

3 a 6.5 feet **b** Yes, minimum is 2.5 feet.

4 5 feet; the minimum from the table is 10 feet but this would only extend 3.5 feet above roof. Minimum height table says 5 feet.

5 a 2 sq in **b** 5 sq in **c** 13 sq in **d** 16 sq in

Task 5 (extension)

1 a $H = 8$ feet, $L = 5$ feet; fan: minimum 83 to maximum 473; natural 313

b $H = 10$ feet, $L = 2$ feet; fan minimum 26 to maximum 289; natural 195

2 The pipe needs to be 8 feet from vertical wall so need a lateral distance of 5 feet minimum. This gives a minimum height to the roof of 9 feet. This would be OK as even an 8 foot high 6 inch pipe with 5 feet laterally has a maximum of 173 000 Btu. Pitch is 12/12, which means at least 4 feet above the roof.

Alcohol

Task 1

1 Students' answers will vary, but could include: drink was cheaper because it was not taxed; there were more pubs then (1 for every 300 people); there was little other entertainment.

2 1950   **3** 12–13 litres   **4** About 60   **5** About 9.5

Task 2

1 a 15 **b** 40 **c** 1.375 **d** 1.6 **e** 2.3

2 a 54% **b** 37.5% **c** 11% **d** 8.3% **e** 1.8%

3 2.4   **4** 1.8   **5** 19.7

Task 3

1 She would have had 2.28 units, so will probably be over the limit.

2 Just about; she consumes 2.75 units, which would put her blood alcohol at under 9%, and she loses 3% during the course of the meal.

3 He consumes 8.6 units, which puts his blood alcohol at about 21%, but he loses about 4.5% during the 3 hours so his blood alcohol will be about 16.5%.

4 They each have about 3.75 units so the girls will have about 16% and the boys about 11%.

5 He consumes 5 units so he has a blood level of 12% and loses about 2% during the meal. He needs to lose another 2% so he needs to walk for 1 hour 20 minutes and even then he will be on the limit so probably should not drive.

Saving energy

Task 1

This answer is approximate and answers could be different but should be about the same.

- Area bedroom 1: just less than 12 m^2 so 2400 Btu per hour
- Area bedroom 2: 8.75 m^2 so 2100 Btu per hour
- Area study: 6.25 m^2 so 1800 Btu per hour
- Area bathroom: about 6 m^2 so 1800 Btu per hour
- Area dining room: 6 m^2 so 1800 Btu per hour
- Area kitchen: 8.75 m^2 so 2100 Btu per hour
- Area lounge: 16.5 m^2 so 3000 Btu per hour

Total: **15 000** Btu per hour

Heating is on for $5 \times (2 + 6.5) + 2 \times 16.5$ hours per week = 75.5 hours per week;

75.5 hours per week ≈ 3926 hours per year; $3926 \times 15\,000 = 59\,000\,000$ Btu per year,

59 000 000 ≈ 60 million ≈ 40×60 kg CO_2 = 2.4 tonnes per year (1.5 million Btus ≈ 60 kg of CO_2 gas)

Task 2

This will vary according to each student or group.

Task 3 (extension)

There are many new cars that are both fuel-efficient and low carbon emitters. The road tax payable is now dependent on the emissions and for some cars the road tax is zero.

Rugby numbers

Task 1

Melrose: $(7 \times 5) + (5 \times 2) = 45$

Hawick: $(3 \times 5) + (1 \times 2) + (2 \times 3) = 23$

There are other ways of scoring 23 points, such as: 4 tries + 1 penalty; 2 tries + 2 conversions + 3 penalties; etc.

Task 3

Try worth 4 points:

Melrose: $(7 \times 4) + (5 \times 2) = 38$

Hawick: $(3 \times 4) + (1 \times 2) + (2 \times 3) = 20$

Try worth 3 points:

Melrose: $(7 \times 3) + (5 \times 2) = 31$

Hawick: $(3 \times 3) + (1 \times 2) + (2 \times 3) = 17$

Bike race

Task 1

Stefan Nimke won in a time of 1:00.793s

Task 2

The medal winners were: Stefan Nimke (1st); Teun Mulder (2nd); François Pervis (3rd).

Body mass index

Task 1

Her BMI is 16.37. She is underweight.

Task 2

Height 1.80 m, weight 81 kg

Height 1.85 m, weight 85.5 kg

Task 3

Height 1.75 m, weight 53 kg, BMI = 17.3: underweight

Height 1.20 m, weight 31 kg, BMI = 21.5: healthy weight

Height 1.65 m, weight 57 kg, BMI = 20.9: healthy weight

Height 1.60 m, weight 80 kg, BMI = 31.3: obese

Task 5

Imperial units have been used instead of metric units. Convert the Imperial measurements to metric: 6 ft = 1.83 m, 15 stone = 95.25 kg; BMI = 28.4. Dad is overweight.

Task 6

BMI = 34.1

BMI is not a very good indicator for fit people and sportspeople, because muscle weighs more than fat, so someone with a lot of muscle can have a high BMI.

Green travel

Task 3

London – Edinburgh by car: varies depending on the type of car used.

London – Edinburgh by train: $661 \times 0.0602 = 39.79$ kg

London – Edinburgh by coach: $661 \times 0.029 = 19.17$ kg

London – Edinburgh by air: $661 \times 0.1753 = 115.87$ kg